THE FACE OF AUSTRALIA

By the same author

ANCIENT AUSTRALIA
SOUTH WITH MAWSON

CHARLES FRANCIS LASERON
Revised by J. N. Jennings

THE
FACE
OF
AUSTRALIA

THE SHAPING OF A CONTINENT

ANGUS AND ROBERTSON

ANGUS AND ROBERTSON (PUBLISHERS) PTY LTD
102 Glover Street, Cremorne, Sydney
2 Fisher Street, London
107 Elizabeth Street, Melbourne
167 Queen Street, Brisbane
89 Anson Road, Singapore

First published in 1953
Second edition (revised) 1954
Reprinted 1957, 1961, 1964, 1965
Third edition (revised) 1972

© *Estate of Charles Francis Laseron* 1972

National Library of Australia
card number and ISBN 0 207 124094

Registered in Australia for transmission by post as a book
PRINTED IN AUSTRALIA BY WATSON FERGUSON & COMPANY, BRISBANE

Acknowledgments

Though this revision of Laseron's book has purposely been of a limited nature for reasons set out in the *Appreciation*, it has certainly been intended to make its contents accurate in terms of present knowledge. There is a good deal of material in this book about plants; in this regard I sought the help of Mr G. S. Hope of the Department of Biogeography and Geomorphology, Australian National University, and I am grateful for the care with which he commented on this side of the book.

The sources on which many of the figures depend are indicated in the captions. I must give particular thanks to Mr E. B. Joyce, Department of Geology, Melbourne University, who lent a copy of an unpublished map for the preparation of the up-to-date map of the volcanoes of western Victoria. In redrawing more than half of the figures to take account of fresh work and better maps, I found that many of the place names had changed officially and topographic heights had been revised. Therefore I asked Mr K. Mitchell, Department of Human Geography, Australian National University, who did the fair drawing from my sketches and also redrew the remaining original figures to suit the new format and style of this book, to check the other place names and altitudes. This book has benefited greatly both from his professional cartography and this help with the related side of the text. A fresh set of photographic prints had to be obtained for this edition and once again many Federal and State organizations co-operated most willingly. These authorities and certain generous individuals who provided pictures are named in the captions to the photographs but it is here one can thank them for their courtesy and generosity.

J. N. JENNINGS
ANU, Canberra, 1971

An Appreciation

Not many books popularizing a branch of science are still in demand nearly twenty years after their first appearance, but this is true of Charles Laseron's *The Face of Australia.* It is not surprising, of course, that there should be demand for books which attempt to explain the scenery that catches the eye while one is travelling through this country. Most people have cars and are using them to travel farther and farther afield. Air travel, almost equally universal now, provides different and thought-provoking views of features familiar on the ground; from the air we are better able to grasp their relationships to one another as well. Apart from young people growing up who would find such a book as this useful in order to learn something of their native country, there has been throughout the two decades since this book's first publication a continuous influx of new settlers to whom the very strangeness of this land, in contrast to the Northern Hemisphere scenery, stimulates interest in its whys and wherefores. Moreover there is more than a trickle of foreign tourists nowadays to add to the call for books that satisfy this interest.

It is true that there have recently been published for the Australian audience some excellent, up-to-date introductory textbooks on geomorphology, a science which can be defined as the study of the nature, distribution, origins and evolution of the forms of the earth's solid crust and of the natural processes that fashion them. But, however well done, textbooks with their formal structure, their constraint within the bounds of disciplines, and their customary verbal economy will not appeal to many of those with a purely leisure-time or cultural involvement with the scientific study of scenery; or at least will not appeal until their intellect has been imaginatively stimulated by a book like *The Face of Australia,* which is about as far removed in its whole style from a textbook as any work still fundamentally scientific in purpose can be. This makes for its wide appeal.

Laseron is far from restrictive about what he includes. Though at the very beginning he makes his central purpose clear with a familiar quotation from Tennyson's *In Memoriam,* he happily dilutes his geomorphology with snippets about explorers' journeys, aboriginal legends, the economic use of the land and its resources, together with larger interpolations, indeed whole chapters, on vegetation and its history. Of course, vegetation plays a vital role in the shaping of the land through its effects on the processes of erosion and deposition, but this criterion does not govern what Laseron included about plants. Laseron was, in fact, one of the last of a dying breed—the natural historians. For too many people today natural history conveys a rather sentimental interest in furry animals. In the heroic days of Australian field science, men like Ralph Tate, Professor of Natural Philosophy at Adelaide University, were concerned with all natural realms; rocks, landforms, soils, the weather, plants and animals were all grist to their mills.

Though Laseron's scientific writings proper were only in the geological and zoological domains, he was much involved with plants as well as rocks and animals in his first post as Collector for the Sydney Technological Museum to which he was appointed in 1909. While there he helped with important research on eucalyptus oils conducted at that institution. All natural things enlivened his curiosity.

Laseron's interests extended beyond natural history. After the First World War until 1929 when he retired from the Museum he was Officer in Charge of Applied Art there, and his writings on ceramics and objects of art reveal intellectual engagement in this direction. Such catholicity of interest in the author matches admirably the attitude of mind of much of the audience for a book of this kind. To have concentrated rigidly and intensely on the skeleton of the landscape—the sculpture of the solid crust—when natural scenery in the broad sense is compounded of so much else besides would have betokened an insensitive and ineffective scholasticism quite foreign to Laseron's nature.

More important still was Laseron's great ability to communicate to a wide audience. His papers in the scientific journals exhibit the complete appurtenance of the scholar, but he was able to set all this aside and to write simply and evocatively without batteries of abstruse terms. Indeed, one of the remarkable things about *The Face of Australia* is the mere handful of scientific words employed to discuss a wide range of problems, some of them of a rather complicated nature. No doubt his personal history as well as his character inspired an inner motivation which can so often lead to a mastery of the appropriate means. He was a modest man by nature, and his father's equally modest income as an Anglican parson prevented him from going to university though his abilities clearly warranted this. Laseron retained a sympathy for those endeavouring to improve their knowledge under difficulties. He therefore found it easier than many to convey the essentials of the subject in leisurely description and with homely analogy.

Lastly, but inseparable from the previous attributes, the book itself is redolent of the man. The impersonality of the scholar is cast aside; warmth of feeling enlivens fact and argument throughout. "Never a rich man, he nevertheless enjoyed a rich life," wrote D. F. McMichael and G. P. Whitley of him in an obituary notice. An important element in that richness was an unalloyed enjoyment of the beauty of landscape and pleasure in unravelling some of its workings. There is an attractive freshness of imaginative response to the force and persistence of natural agencies attacking the rocks and perpetually recasting the shape of the land.

It is clear from references in this book to early field trips that Laseron had this perceptive appreciation of natural processes from the beginning of his studies, but it must have been reinforced by his grim experience in the "Home of the Blizzards"—Adelie Land—on Sir Douglas Mawson's 1911-13 Antarctic expedition when he was still a young man. These experiences he recounted much later in his book *South with Mawson*. *The Face of Australia* was written later still, in retirement after the Second World War, when heart trouble set limits to his physical activities. Thus there is a very real flavour of recollections in tranquillity of the scientific travels which had enriched his life.

Patriotic feeling stands forth frankly on many pages as might be expected of a man who landed at Gallipoli in 1915, was wounded there, and wrote one of the first accounts of the landings to be published in Australia. But the honest love of his country that pervades this book is totally free from jingoism and did not warp his judgement. There are no chimeric claims here of the greatest in the world, no categorical superlatives, not even any rash biggests in the Southern Hemisphere which usually reveal a blissful ignorance of the continent of South America. Old-fashioned though some of the patriotic expressions may be in style, they nevertheless strike a modern and most apposite note because they are always part of the pleas for conservation of flora, fauna, and scenic monuments. Indeed there can be no doubt that the previous editions of this book which have reached thousands of readers over the last two decades have helped to promote the growing opinion that we must preserve adequate national parks and reserves of land in a natural condition as habitats for the country's wealth of organisms and as spiritual lungs for the denizens of our swollen cities.

With such qualities in *The Face of Australia* there is good reason to continue to make it available to new readers. But this was not a simple proposition, for a new edition could have taken the form of a straight reprint, a complete revision, or something in between. Since this book was first published a great deal of research into the nature and development of Australian landforms has been carried out by geographers, geologists and soil scientists in universities, government departments and instrumentalities such as the CSIRO. Much remains to be found out; indeed often enough more detailed work has cast doubt on the initial analyses of landforms by the pioneers without determining a sure explanation. Nevertheless a great deal of fresh knowledge has changed the background. Moreover, where fresh geomorphological studies are lacking, general geological mapping has often had some effect on what we must think about many landforms. A vast amount of fresh topographical mapping has rendered the book wrong on many points of factual detail.

So it was not satisfactory simply to reprint. The book as it stood was not a historical document or a piece of Australiana; it could still be read in order to learn about Australian scenery and deservedly so. On the other hand, to make it generally indicative of the present state of knowledge of our landforms, there would have had to have been so much addition as well as correction to the text that there would have been a grave danger of the book completely losing so much of its essential character as really to have ceased being Laseron's book. So personal in approach and style is it, a reviser would have found it practically impossible to match interpolations and modifications with the original text.

Therefore the sound course seemed to be to adopt an approach between these two extremes, and this is what I have done. The original text can be relied on to give a stimulating introduction to the subject, since Laseron's choice of matter was quite reasonably representative. But wherever the text was incorrect or misleading about present thought on a particular subject it has been replaced or supplemented. Rather than full revision in the light of everything we know now, the basis of this new edition is therefore amendment of the original contents to make it possible to read the book with confidence that what is learnt is not antiquated but effective knowledge.

J.N.J.

Contents

List of Text Figures

Illustrations

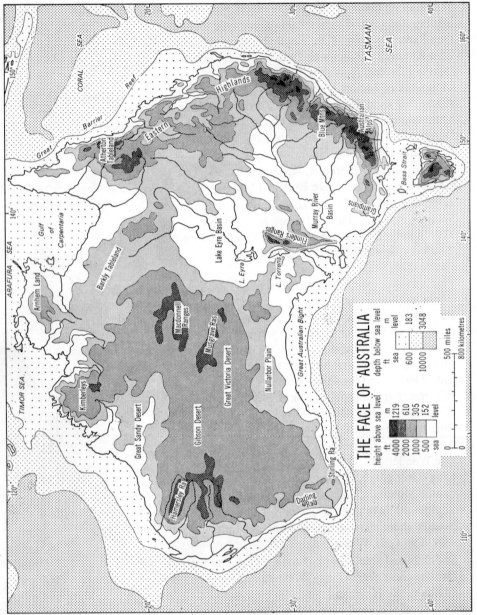

1. Relief map of Australia.

The Australian Landscape

The hills are shadows, and they flow
From form to form and nothing stands.
They melt like mists, the solid lands,
Like clouds they shape themselves and go.

When Tennyson wrote those lines he used no poetic licence but spoke the literal truth. The world is an old, old place. Its story has been traced back through the rocks for some 3,000 million years. It is a story of endless change, of mountains rising and disappearing, of land changing to sea and sea to land, of alternating frigid and tropical climates, of periods of intense volcanic activity and others of quiet stability, of arid regions which have become fertile, of fertile regions which have become desert.

Spectacular scenery is the attribute of youth. Lofty snow-covered peaks reaching above the clouds, the glaciers which stream down their sides, the lakes hidden in the valleys, rushing waterfalls, great gorges which split the plateaux, spouting geysers, the craters of active volcanoes—these would seem to be eternal. Actually they are all young, the offspring of the very latest of the many cataclysms which have convulsed the surface of the earth in past ages. And like others which have preceded them, they are ephemeral, fated to disappear in the processes of unceasing dissolution.

Much of Australian scenery is not young, and thus, by oversea standards, not spectacular. Some of it is beautiful, much of it is distinctive, all of it is interesting. To the early settlers, separated from their homelands by the width of the world, nostalgia oft-times blunted perception to the beauty of the unfamiliar. More understandable were the impressions of the very first explorers to see the shores of *Terra Australis,* the legendary land of the south. The arid north-west coast, viewed by Dutch seamen and then by our own Dampier, presented the nadir of inhospitality, and well justified the initial view that here was a land unlikely ever to be of use to mankind. Even well after the first settlement on the east coast these views persisted. Journals of the early settlers abound with unflattering references to the new colony. In Barron Field's book on New South Wales, in that part entitled *First Fruits of Australian Poetry*, appears the "Ode to the Kangaroo", which commences

"O kangaroo, O kangaroo,
Thou spirit of Australia,
Which redeems from utter failure
This land of desolation."

This epitomized the feelings of many of the early visitors to our shores. The famous Charles Darwin, when he came to Australia in the *Beagle* in 1836, though impressed by the progress of the young colony, saw little future for it save in oversea commerce and possible manufacture. As he left he wrote in his journal, "Farewell, Australia! You are a rising child, and doubtless some day will reign a great princess in the South; but you are too great and ambitious for affection, yet not great enough for respect. I leave your shores without sorrow or regret."

To those of us who have put down roots in the soil, what the pioneers found unattractive now binds us to the country. The unfamiliar has become the familiar; it has become part of us and we of it; it is the spirit of our national pride, the basis of our love of country.

Whereas in so many lands scenic beauty is an attribute of youth, in Australia it is expressive of a great age. Parts of it are amongst the oldest land surfaces in the world. Its contours speak of vast antiquity, of long eras of stability, of freedom from the cataclysms that are still changing the shape of other countries. Even the animals and plants, developed in long isolation from the outside world, are different.

It is a long time since geological events of catastrophic magnitude took place in Australia. It was before the Permian Period, that is 230 million years ago, that alpine ranges with peaks above the snow line were last convulsed upwards by the forces of nature. Since then earth movements have been on a diminishing scale. Volcanoes have at times poured out devastating floods of lava and incandescent ash, there have been up and down movements of the earth's crust and some modest compression to produce low ranges at the extreme eastern and western fringes, but nowhere in Australia has there been mountain building on a grand scale. While Alps and Himalayas and Rocky Mountains and Andes have been, and are being, thrust into the sky by intense pressure and folding, Australia has slept on serene and untroubled.

Only occasionally in the last few million years have there been some uneasy stirrings of old mother earth, stirrings sufficient to leave some imprint on the surface of the land. There have been gentle risings and fallings, like the rhythmic breathing of a sleeping giant. To these we owe such high land as at present exists. This consists of tablelands rising at the most to a few thousand feet above the level of the sea and some isolated ranges that are the remnants of once active volcanoes.

Even these are in process of dissolution by the ever active forces of nature, rain, frost and the wind. Should nothing interrupt the sequence of events, should there be no future movements of the earth to build more high land, the result is inevitable. The surface of the hills will become lower and lower until the land is reduced to one vast plain slightly above the level of the sea. When the rivers flow so slowly that they can no longer carry their burden of silt to the ocean, destruction will cease and equilibrium be reached.

Although that stage is still far distant, Australia is even now a land of low relief. It is a country of great distances. In such an area there is naturally a diversity of scenery: there are tablelands dissected by great gorges, great plains both on the coast and in the interior, even great deserts. There are few lakes and no mountains above the line of perpetual snow. On the whole there is uniformity from north to south, from east to west, but that uniformity is not monotonous, since there is considerable difference in detail.

The uniformity of Australia facilitates its presentation as a whole. The continent is not only a geographical unit, it is a biological unit also, and at its present stage in the scheme of things a human unit as well. Whereas in Europe it is possible in a day's journey to pass through many different countries, with different scenery, different languages, different customs, a much longer journey in Australia may take the traveller to almost a replica of the place he left. We Australians find this comforting, though the new-comer may complain of monotony. The Western Australian visiting New South Wales, the Tasmanian visiting Queensland, may find some difference in climate or scenery, but little greater than if he travelled a much shorter distance within his own State. He will be at home at once. The forests may be of different species of eucalypt, but they will still be "gum" forests. If the country is high, the ravines will present the same aspect; if the country is granite, he may expect the same rounded boulder-covered hills, with familiar farms nestling at their bases. Only in some places, in the tropical jungles of northern Queensland, amongst the lakes of central Tasmania and on its wild west coast, amid the red mountains of central Australia or the sand dunes of the central deserts, will there be great diversity. But the people will be the same, speaking the same language, thinking the same thoughts, with the same outlook on life, eating the same kind of food and drinking the same drinks.

Australia is thus knitted into a coherent whole. A European may be an Englishman, a Russian or an Italian, but an Australian is an Australian, whether he comes from Victoria or the Northern Territory, and whether his forebears came from Scotland or from Germany. In less than two centuries a nation has risen to maturity, a nation which from geographical environment alone has become probably more coherent and indivisible than any other on earth. The average Australian, even if a city dweller, has among other things acquired a love of the bush, a love of space in which to live. Above all he has learned to appreciate freedom, freedom to live his own life, think his own thoughts and speak his own words. All is bound irrevocably with the scenery amongst which he lives.

Apart from its spaciousness, it is the gum tree which gives the Australian landscape much of its character. The giant gums of Tasmania and Gippsland, the karri and jarrah forests of Western Australia, the twisted white gums of the sandstone tablelands, the drooping red gums of the inland rivers, the stringybark, ironbark and box of the open forests, the tallowwood and spotted gum of the coast, the mallee scrubs of Victoria—all are of the genus *Eucalyptus* which dominates the Australian bush, providing the strongest note in nearly every scene. The gum tree may lack the luscious massive greenery of the deciduous trees of Europe or the symmetry of the pines and firs, its lines may be irregular, and its vertical leaves may give little shade;

but the sunlight filtering from above weaves patterns of light and shade on its trunks and branches and gives it a beauty all its own. This beauty Hans Heysen and other painters have caught and given to the world.

Of all countries the primitive areas of Australia are the safest for mankind. Natural hazards are at a minimum. Possible hunger and thirst and the fear of getting lost are the greatest. No traveller need carry firearms for protection. There are crocodiles in the rivers of the far north, but no predatory creatures roam the plains or forests, or spring from the trees upon the unwary. There are poisonous snakes, but they are timid creatures and easily avoided. The funnel-web and the red-backed spider are occasional minor hazards, and there is some discomfort possible from mosquitoes, sand flies, or, in the coastal scrubs, from stinging trees, leeches and ticks; but disease bearing insects are conspicuously absent. There are, for example, no deadly typhus-bearing ticks, or flies that carry the dreaded sleeping sickness. The sea has a few minor dangers, man-eating sharks, a jelly-fish that is to be dreaded, and, on the Queensland coast, a sea-shell whose sting may be fatal. In spite of all these, the Australian countryside is a remarkably safe place.

The Australian is fortunate that close to the main cities there still lie areas of virgin bush where it is possible to find direct contact with nature. There are few who have not boiled a billy under the gum trees. Within 60 miles of Hobart lie the rugged wilds of western Tasmania, much unexplored and until recently quite inaccessible. Melbourne has the Dandenong Ranges close by, Adelaide the Mount Lofty Ranges, Perth the Darling Ranges, Sydney the National Park and the Blue Mountains, Brisbane the rain forests of the Lamington Park. Apart from the vast arid regions of central Australia there are even now many places quite close to habitation where man has rarely if ever trod. The gorges at the headwaters of many of the coastal rivers have seldom been penetrated, save by lone prospectors and seekers after cedar and other timbers.

It is in the solitude of these places that the real Australian atmosphere is to be found. When an Australian abroad is homesick, it is not for the cities that he longs. My own nostalgia has risen in the memory of a camp by a rushing stream at the bottom of a gorge where the sun shines but a few hours each day. There the smoke of the camp fire was redolent with the scent of burning gum leaves; there was the splash of a lizard taking to the water, the twittering of birds in the trees, the laugh of the kookaburra, the call of the thrush, the crack of the coach-whip bird. The mimicry of the lyre bird echoed across the gully, while wonga pigeons and satin bower birds fed on berries in the neighbouring brush.

Memories may be of a lonely mountain-top in Gippsland, overlooking a vast expanse of hills and gullies, all covered with a primaeval gum forest, with no habitation in sight, and a solitary wedge-tailed eagle hovering overhead. Or they may recall a white beach isolated between towering headlands, or golden sunsets over far-flung plains.

Beneath all—and here perhaps it is the geologist speaking—is the evidence of a remote past. Nature is a tireless sculptor, for ever fashioning new master-pieces, yet, as if unsatisfied, commencing their destruction from the very moment of their completion. Something of the old always remains to be built into the new. The shape of a hill, the contour of a waterfall, the rocky ledges on the sides of a gully, the meanderings of a river, the alluvium of a plain are all links with the past, and within them lies the story of what has gone before.

This is the real meaning of landscape, tracing from the present shape of the earth the history of what has happened during countless millions of years. To look at the hills and valleys of the present day is to visualize the seas and mountains, the lakes and glaciers of bygone periods, and to conjure up the many races of strange plants and animals that have come and vanished, passing like a moving picture across the screen of Time.

By Way of Explanation

In this book technical terms have been discarded as far as possible, but there are some terms which are really necessary, since there are no words in common usage to replace them. All such words are not strictly technical; in fact, most of them appear in ordinary dictionaries, but as dictionary definitions are of necessity very brief, a rather fuller explanation is given here. This, it is hoped, will obviate any misunderstanding of the sense in which they are used.

The Geological Record. When seeking the causes of present scenery it has been frequently necessary to refer to events in the remote past, and to use the names of the different periods into which geologists have divided the earth's history. The discovery that the past may be divided into a number of great epochs or periods broadly coincident throughout the world was made early in the nineteenth century, and since then armies of geologists, zoologists and even physicists have engaged in patient research in correlating the changes in geography and life that took place in different countries. Though much detail remains to be filled in, and though much evidence of the past has been for ever destroyed, the broad results of this research may be taken as substantially correct. We can thus speak almost as definitely of the Permian Period as we can in human history speak of the House of Tudor in England or the Ming Dynasty in China.

It is even possible to give an approximate estimate of the antiquity of the different periods in years. This has been one of the results of the study of radioactivity in the earth. Certain rock minerals change chemically as a result of their radioactivity. Each radioactive change takes place at a fixed rate which has been determined so that the difference in the composition of a mineral from its original one gives a measure of its age. Thus potassium in minerals in rocks such as basalt gives off argon gas in its radioactivity. The amount of this argon relative to potassium enables the time of formation of the rock to be calculated provided no gas has been lost to the air. Rubidium changes to strontium in the course of radioactive change in other minerals and their ratio also gives dates. Through these methods especially, the time-table for the rocks as given here may be accepted as approximately true.

The following table is a summary of what is termed the Geological Record, giving the names of different eras and periods as used in this book, but omitting certain broad groupings, and the many sub-divisions into which the periods are divided. It also gives the approximate time in years from the present when these different periods commenced.

ERA OR PERIOD	YEARS AGO BEGUN
Archaeozoic Era	4,500 million
Proterozoic Era	2,300 „
Nullaginian Period	
Carpentarian Period	
Adelaidean Period	
Cambrian Period	550–510 „
Ordovician Period	480 „
Silurian Period	435 „
Devonian Period	405 „
Carboniferous Period	362 „
Permian Period	280 „
Triassic Period	235–230 „
Jurassic Period	180 „
Cretaceous Period	135 „
Tertiary Period	65 „
Paleocene Epoch	
Eocene Epoch	
Oligocene Epoch	
Miocene Epoch	
Pliocene Epoch	
Quaternary Period	
Pleistocene Epoch	2–3 „
Recent (Holocene) Stage	10,000

Erosion. Everyone knows what erosion means. Indeed, erosion of the soil by rain and wind is one of the pressing economic problems of the present day. Few nevertheless realize what a potent factor it is and has been in shaping the face of the earth. Not only the soil is washed away, but in the course of time even the hardest rocks are broken down and the highest mountains worn away. There is no need to outline the various processes by which this is achieved—such information is given in any book on Geology—but in these pages it will be seen again and again how the shape of present scenery is the result of erosion in the past.

Peneplain. When high land has been exposed to a long cycle of erosion, and has been worn to such a low level that the rivers can no longer carry away their burden of silt, the land surface is termed a peneplain. This is not necessarily an absolute plain, for low residuals of harder rock may remain for long periods. The peneplain virtually marks the end of erosion, and should no further earth movements take place, it may remain as such indefinitely. When, however, the land is again elevated, erosion is renewed, and the only evidence of the former peneplain may be a series of hills all of nearly the same height rising above the general level of the new tableland.

Pediplain. The pediplain is another kind of lowlying surface to which the processes of erosion reduce the land and it is more important in Australia's scenery. The peneplain is chiefly due to the wearing down of slopes; rivers cut steep-sided valleys to begin with but later the valleyside slopes are worn down to lower and lower angles. In contrast, the pediplain is due mainly to a process of wearing back of valleysides from the streams. Valleysides may retain their early steepness, part of the slope retreating at the same angle and leaving in front a gentle foot slope called a pediment. The hills above these pediments remain steep even when they are but small remnants of the original relief.

Faults, fault scarps and faultline scarps. Faults are long fractures in the rocks along which the land on one side has either risen or subsided relative to the other. The displacement may be only a few inches or it may be many thousands of feet. Faults which took place in the early geological periods are often obscured by a covering of later formations, but others of more recent origin have a prominent effect on the landscape. The line of cliffs or heights bordering a sunken area, or alternately on the margin of a risen area, is called a fault scarp.

However, in Australia especially, many scarps along faults are not simply due to the displacement of the land on either side of the fault. The original fault scarp in many cases has been worn down by erosion during the formation of a peneplain or a pediplain. Nevertheless, though the original effect of faulting on the land surface may have disappeared, rocks of very different powers of resistance to erosion may still lie on either side of the faultline. If erosion is resumed, through uplift or other cause, fresh attack on the crust is likely to remove the weaker rock faster than the other, lowering the land on the one side of the faultline more than on the other. Thus a new scarp may form due to erosion but guided by the faulting which had taken place long before. Such a scarp is called a faultline erosion scarp or more briefly a faultline scarp.

Horsts and rift valleys. These terms are closely associated with faults. A horst or block mountain is an area of country which has risen as a whole, and its margins are defined by faults. Horsts which have newly risen are bounded by fault scarps, but older horsts or those which have risen very slowly have been subjected to much erosion, and their scarps are worn and broken. Their origin is then not always apparent. A rift valley is a sunken area, the reverse of a horst, and the fault scarps face inwards towards a central basin. When, as often happens, two great faults run parallel for a considerable distance, the rift valley may be greatly elongated, as for instance the Jordan Valley in Palestine and the Great Rift Valley of central Africa. Some older horsts and rift valleys are bounded by faultline scarps in place of the original fault scarps.

Bedding planes. Coming to the rocks themselves, many, such as sandstones, shales and limestones, have been laid down as sediments on the beds of former seas and lakes. Normally one layer or stratum has been deposited horizontally upon another, and the surface of each of the numerous strata is known as its bedding plane.

Dip and strike. Rock strata do not always lie in their original horizontal positions. The earth movements which elevated them above the sea or subsequent earth movements have often been accompanied by intense lateral pressure by which the strata have been bent or folded. They then become inclined at an angle or may even be quite vertical. The angle at which they are inclined from the horizontal is known as their dip, whilst the direction of a horizontal line in the bedding planes of the rocks is called the strike. The greatest or true dip is at right angles to the line of strike. When the inclined beds have been at a later stage eroded, their edges will be exposed on the surface, and the strike of an individual stratum may sometimes be followed for a considerable distance. The dip and strike of rocks and their relative hardness naturally have a large effect on the present landscape.

Sedimentary rocks. Of the common sedimentary rocks, sandstone is familiar to everybody. Sandstone is simply consolidated sand, and when pebbles are included it becomes conglomerate. Shale is consolidated mud, while limestone is composed of carbonate of lime, either deposited chemically from water, or more frequently formed of the remains of shells, corals and other marine organisms. A very interesting rock is varve or varve shale. This is very fine material, mainly glacial mud, carried away by rivers and redistributed on the bottom of lakes or the sea. It is finely laminated, and each lamina consists of slightly coarser sediment at its bottom than at its top where it is cut across sharply by the next lamina. The coarser sediment is deposited in summer when the water is turbulent through ice melting and river floods. The succeeding finer sediment accumulates during the winter when the rivers are silenced by freezing and, beneath the frozen surface of the lake, fine clays fall quietly through calm water onto the bottom. So by counting these laminae a useful timetable is obtained for measuring events during glacial periods.

Intrusions. At a depth within the earth's surface the materials of which rocks are made are extremely hot, and under such tremendous pressure that they are in a plastic condition. When the pressure is temporarily eased from various causes, the plastic material becomes molten and forces its way upward through the overlying strata. It may cool and solidify without reaching the surface, or it may flow from a volcanic vent as lava. Granite and similar rocks are intrusions which have solidified at considerable depths, and are only exposed on the surface by later erosion. Nearer the surface, intrusions take place as thin bodies which are distinguished chiefly by the direction in which the molten rock was forced. Dykes are narrow intrusions, which penetrated other rocks more or less vertically upwards. But the line of least resistance may have been sideways. The horizontal layers thus formed are called sills if the molten rock has spread along bedding planes, and sheets if they have cut across the beds. All have played their part in shaping the present scenery.

Granite and allied rocks. Rocks which have solidified from molten matter deep within the earth are generally coarsely crystalline in texture. The most familiar is granite, grey or red according to its composition, but always with an excess of the substance silica, which is present as the mineral quartz. Similar in structure to granite are dark-coloured rocks containing less silica and with no quartz present, but varying in composition and composed of different minerals. Of these, scenically important are dolerite, a dense, dark rock, finer in structure than granite, and gabbro, very coarse in texture and black or bronze-black in colour.

Smaller intrusions, solidifying at shallower depths and cooling more rapidly, are often composed of one of the many varieties of porphyry. These are finer in texture than granite and have a rather different crystalline structure. They are very hard and resistant, and produce much rugged and picturesque scenery.

Lava. The term lava includes a great variety of rocks, but for our purpose two groups may be recognized. Lavas are very fine in texture, with small crystals set in a base of natural glass. Basalts are very common rocks; they are dark in colour and are liable to decompose rapidly into rich red soils. Rhyolite and trachyte are lighter in colour, harder, and do not decompose so rapidly.

Pyroclastic rocks. Volcanic eruptions are often accompanied by violent explosions. Liquid matter within the neck of a volcano, as well as the rocks through which it erupts, are shattered and hurled into the air, the heavier masses falling round the vent and building up the cone of the volcano, while the lighter material may be scattered over the surrounding area for great distances. These fragmental volcanic rocks are distinguished according to their size. The coarsest form agglomerate, the intermediate sized material is scoriae, whilst the finest is called tuff.

Metamorphic rocks. Many rocks, particularly those belonging to the early periods, have since become altered both in texture and chemical composition. Such alteration is due mainly to great pressure and heat. Sandstone becomes altered to hard compact quartzite, and shale subjected to pressure turns into slate. When the pressure and heat are very great, rocks such as shales and slates become partially recrystallized and turn into the foliated rock called schist. Other rocks, such as granite are altered to more coarsely foliated rock along with the formation of new minerals including garnet. This is known as gneiss, and many of the oldest rocks of the world are of this nature. Amongst other rocks changed, limestone has been recrystallized into marble, and gabbros by chemical alteration have been converted to the rather soapy green rock serpentine.

Moraines. Glaciers have played a large part in moulding modern topography. Glaciers are rivers of ice, the consolidated snow moving by its own weight from the mountains to lower levels. The greatest glaciers of all are such as still cover Greenland and Antarctica in vast ice sheets many thousands of feet in thickness. In the past similar ice sheets covered other parts of the world, and they have left abundant evidence of their passing. In favourable circumstances glaciers are capable of great erosion and gouge out deep valleys, though if they override mountains, they round them rather than remove them. On the other hand in other conditions glaciers only touch up the landscape, doing less erosion than if rivers were free to attack the ice-covered surface. They are also great carriers, bearing on their surface or pushing beneath them masses of rock and clay, the debris from the rocks over which they pass. When the ice finally melts this material is left behind and is known as a moraine; the consolidated mixture of rock and clay is the rock tillite. Large individual rocks which have been transported by glaciers are known as erratics, and when the glaciers reach the sea and break away as icebergs, erratics may be carried by floating ice and eventually deposited on the sea bed hundreds of miles from their source.

Cirques. Cirques are great semi-circular amphitheatre-like valleys scooped out by glaciers, particularly on the margins of tablelands, or where glaciers have formed near the summits of high and steep mountains. Such glaciers are generally wide in proportion to their length and have a very steep fall.

Cols. A col is more descriptively known as a saddle, and is a ridge joining two higher peaks or higher parts of a range. It is itself higher than the surrounding country. From col comes the term geocol, which is really a col on a grand scale, in itself an elevated area of country of considerable size but lying between even higher country on either side. In the true geocol there should be some definite geological connection between its structure and that of the adjoining higher land.

Rivers. Most of the terms referring to rivers are well understood, but there are one or two which are popularly misapplied. Thus the area drained by a stream or a river is its basin and not the watershed as some people call it. A watershed is really the same as a divide, that is the line dividing two river basins or river systems. It may, but does not necessarily, follow the line of the highest ground; in fact, the main divide between the eastern and western waters of eastern Australia is often little more than a ridge and well away from the highest parts of the tableland. The estuary of a river is not merely the wide mouth of a river; it is that part of the mouth of a river affected by the tides. It may in fact go far inland from the actual mouth, the water is generally brackish, and in short coastal rivers it is very often the limit of navigation. A watercourse is not necessarily a perennially running river or stream; it is the lowest part of a valley or even a plain, that level where water would run if it were there. Many watercourses are nearly always dry, and run only after very heavy occasional rains.

Eustatic. This is the last of the terms we need. Eustatic movements are alterations of sea level as distinct from the elevation and subsidence of the land. As the sea is a fluid body, such movements are naturally world-wide, and they have had considerable effect in modifying the topography of coasts and coastal regions. They may be caused in several ways. In the various glacial periods which the world has undergone, the formation of large ice sheets has withdrawn so much water from the sea that its level has been appreciably lowered. When the ice sheets have melted with the return of warmer conditions, the sea has again returned to its former level. Movements of this kind have been more than 400 feet up and down.

Local land movements also affect the level of the sea in other localities. The sinking of a sea bed or of a land mass beneath the sea increases the holding capacity of the ocean basins, and the level of the sea is everywhere lowered. Similarly the rising of the bed of a sea or its shallowing through the deposition of sediments restricts the ocean basins, causes sea level to rise, and spills the water over the margins of distant lands.

The Bush Comes to Australia

The Australian bush is so integral a part of Australian scenery, it is so near and so interwoven with our pattern of life, that we are apt to take it for granted. It is sufficient to know that it is there, and while we may admire the beauty of trees and flowers, we seldom pause to wonder whence they came and when. What is it that has given the bush its own peculiar character? How does it resemble and how does it differ from the forests of other lands? What has led to these resemblances and differences? These questions are not easy to answer, for while much has been discovered far more is still unknown. The whole story will perhaps never be told. Most of the evidence lies in the past, and of this much has been destroyed and for ever lost.

It is curious that in some ways it is easier to decipher the early part of the story than the later chapters. This is because the first known plants were comparatively lowly and of simple organization. They are thus easier to recognize when found as fossils buried in the rocks; similar types persisted through long ages and changed very slowly, and were thus able to spread widely throughout the world. When in later ages plants became more complex, more specialized to live within limited environments, when for identification it becomes necessary to depend on soft, destructible fruits and delicate, transient flowers, the problem becomes increasingly difficult. This is so because when attempting to solve the origin and development of plant and animal life in any country it is necessary to seek for fossils, that is the remains of things which lived in former ages, and which have been buried and preserved in soft mud and sand now hardened into rock. Such fossils have been found in abundance in Australia as in other parts of the world, and they belong to many different ages or periods in the world's history.

The early part of the story is universally much the same; there is very little difference between Australian fossil plants and those of Africa, Asia, Europe and America. There was life on the earth a very long time ago, and though there is no glimpse of its absolute beginnings, we assume that the first living thing was either some simple, unicellular plant or some minute organism from which both plants and animals developed. The earliest life was almost certainly in the sea and was confined there for long aeons before the land became populated. The oldest fossils are bacteria found in Archaeozoic rocks and living more than 2,400 million years ago. They were followed in the early Proterozoic by small algae such as *Collenia*. Animal life came after the plants, for by their metabolism, that is the chemistry of nutrition, plants can exist without animals, but all animal life is dependent, directly or indirectly, upon the plants.

The land surface of the earth was bare and devoid of life through the early geological periods, and when life came out of the sea to populate the land, the plants came first, destined to provide food for the animals which followed. The first land plants were very simple types, akin somewhat to the seaweeds, somewhat to the mosses. We can picture them first finding a footing between tide marks, where they would be exposed to the air twice a day when the tide was at its ebb. As they became more and more adapted to live out of the water they would creep at first into swamps and marshes and finally to entirely dry land. Fossil plants of this type, thought to be the first land plants in the world, have been found in Victoria in rocks of the Silurian Period, that is upwards of 400 million years ago. It is from such humble beginnings that not only the Australian bush but all other floras originated.

For long ages after this, plants were still of very simple types. Ferns were amongst the earliest; then larger and more complex plants began to develop. The first forests were little more than thickets composed of curious plants called lycopods or club mosses, with straight stems a few inches in thickness. Short narrow leaves grew on the stems and left rhomboidal scars when they fell. By the Carboniferous Period the forests had grown in size, and there were many large trees, but still of simple non-flowering types. It was in the Carboniferous Period that the main European and American coal seams were laid down. In the Permian Period which followed large pine-like trees appeared, and then came the true pines, which for a long time dominated the forests. With the pines and becoming more and more abundant were the cycads, palm-like plants of which our burrawang or *Macrozamia* is a survival. The pines and the cycads up to about 80 to 90 million years ago were the predominant vegetation.

About this time true flowering plants or angiosperms began to appear, and they spread rapidly over the whole earth. The story becomes difficult here, for though leaves of undoubted flowering plants are common as fossils, it is often impossible to further classify them by the leaves alone. Botanists rely upon the flowers and fruits to distinguish the different families, and these are generally so fragile and destructible that they are rarely preserved as fossils. Fossil wood is quite common and sometimes beautifully preserved, but even microscopic examination is insufficient to determine the species and even the family of the original tree, though that of the pines as a whole is generally distinctive.

In spite of this botanists have named many of the earlier flowering plants of Europe and America, and if the determinations are correct, they all belong to existing orders. Amongst the earliest are figs (*Ficus*), the sassafras, magnolias and the flame-tree (*Sterculia*). Slightly later came beeches (*Fagus*), birches (*Betula*), oaks (*Quercus*), walnuts (*Juglans*), tamarisks, maples, laurels and oleanders. Palms were particularly abundant and in great variety. There seems at this time to have been little difference between the plants of America and those of Europe and Asia.

The diversity and high development of the early flowering plants is of great significance. If comparison be made with the animal kingdom, the true flowering plants bear much the same relation to non-flowering plants as vertebrates do to animals without backbones. The various classes of vertebrate animals appeared one by one over many periods, first fishes, then amphibians, then reptiles, birds and finally mammals. The mammals themselves developed slowly, at first such lowly forms as monotremes and marsupials, then higher and higher types, until there came the primates and finally man. The flowering plants in contrast evolved with great rapidity, so that within one period of their appearance upon the earth, many if not most of the existing orders were well established. Some living orders of plants have never been found as fossils, but it does not follow that they did not exist in the past. Such plants as orchids for instance, which are classified by their delicate and intricate flowers, have very little chance of being preserved as fossils, and even if so found, would be difficult if not impossible to recognize.

If it be accepted that most of the existing orders of plants were well established throughout the world by the end of the Cretaceous Period, their subsequent development would be by changes within the orders themselves. According to soil and climate some members of the one order would become adapted to live in moist jungles in the tropics, others on river flats in temperate climates, others on bleak and exposed mountain tops, others on arid sand dunes, others even in the heart of the desert. There would be and indeed is a great deal of parallel development. Plants of quite different orders living under similar conditions have converged in their superficial characters, particularly in their habit of growth and the form of leaves. On the other hand plants of the same genus or family living under different conditions have diverged in the same characters far from each other.

No group of plants bears this out better than the legumes or bean-bearing plants to which our acacias belong. The bean bearers are found everywhere from the jungle to the desert, and species will be found which in leaves, growth and general appearance match many plants of quite different orders, simply because they have adapted themselves to similar environments. The same may be said of many other plants in the Australian bush.

It is now thought that Australia remained part of a much larger continent of Gondwanaland into the Cretaceous Period when the great development of flowering plants took place. Many early arrivals must have migrated overland to this country from the west and south; unfortunately, little is known about them. In the Cretaceous rocks of Queensland great quantities of fossil leaves have been found, but very little research has been done upon them, and the identity of most is unknown. Many superficially resemble plants of the northern hemisphere, but they also resemble others still living in the adjacent rain forests, and it is dangerous to draw many conclusions from them. They do show nevertheless that flowering plants were well established in Australia at the time and it is reasonable to assume that these were the original stock from which much of the present Australian flora has developed.

Up to Cretaceous times Australia had been linked by land with Africa, India and Antarctica, but towards the end of this period, that is some 65 million years ago, these land bridges disappeared and Australia became isolated from the rest of the world. Much of New Guinea was indeed submerged beneath the sea in ensuing epochs, that is during the Tertiary Period. Henceforth the Australian bush developed in isolation, just as did the unique marsupial fauna. To quote Bentham, the noted botanist, "The predominant portion appears to be strictly indigenous. Notwithstanding an evident, though very remote connection with Africa, the great mass of purely Australian species must have originated or been differentiated in Australia and never spread far from it."

No doubt some new elements did come in from outside from time to time. These would be from accidental causes, a seed attached to the wing of some migratory bird or carried by driftwood to the shore, even by an insect blown by gales across the sea, or by the wind itself. The story of how plants migrate, how many seeds are adapted for just this purpose, is a fascinating one, and many botanists have devoted themselves to its study. It is too vast a subject to be more than mentioned here, but results may be seen in the flora of many of the oceanic islands. Such islands as the Hawaiian Islands in the Pacific Ocean and St Helena in the Atlantic have never been connected by land with any of the continents, yet they have a large and diverse land flora. The ancestors of this flora must have in the first place crossed the surrounding seas by various means. Some of it has been so long established that it has become considerably specialized and modified into species confined to these islands.

One of the most interesting Australian trees is the beech, an odd anachronism which is probably one of the few unchanged survivors of the original Cretaceous flora. It would be fascinating, if it were possible, to trace in detail its full ancestry. It is a true beech, and though the generic name of the Australian trees has been modified to *Nothofagus,* it is closely akin to the *Fagus* of the northern hemisphere.

The beeches are found mainly in Tasmania, but there is a patch of a different but closely allied species growing in the rain forest in the Lamington National Park on the summit of the Macpherson Range bordering Queensland and New South Wales at an elevation of 3,800 feet. There stands a forest of these trees, many of them very large and of an age computed as upwards of a thousand years. The older trees present a hoary appearance, their trunks covered with moss and many orchids. They are heavily buttressed at the base, and erosion has often lowered the ground beneath them, so that their arched roots stand bare for 10 to 15 feet above the soil, making them look like monstrous, multilegged insects with trees growing from their backs. In New South Wales farther south there is another beech forest growing on the summit of the Barrington Tops.

In Tasmania there are two species of beech, one of which is deciduous. This is a small wiry plant, little more than a shrub, living only on mountain tops and the high tablelands. In summer the masses of bright green leaves, in contrast with the greyer green of the eucalypts, are conspicuous from afar, while in the autumn, before the leaves are shed, they make patches of golden yellow on the sides of the mountains. The other species is an evergreen, forming a large part of the rain forests of the west. Here the trees are large and of tremendous age, their hoary trunks covered with a great variety of mosses and lichens.

There has been considerable discussion on the original introduction of the beech in Australia, and at first sight it appears as something of an alien in the Australian bush. The most reasonable explanation is that it is an actual survivor of the old flora, the flora which existed while Australia was still connected by land bridges with other continents. Beeches have been identified as fossils from the early Cretaceous rocks of Europe and America, and they probably spread until their range was world-wide. Thus they came to Australia with many other plants and ranged far and wide over the whole continent. In the course of time most of the plants were forced to evolve and adapt themselves to new conditions, but the beeches remained constant in their ways and habits, not unlike an Englishman who persists in dressing for dinner in the wilds of Africa. In the struggle for existence they paid the penalty for their conservatism, becoming virtually extinct except in the few localities where conditions are still favourable to their existence.

One of the big factors in the development of the Australian bush has been climate, both past and present. There have been many changes in Australian climate since the Cretaceous Period, some continental or even world-wide, others local. The Cretaceous Period has been considered by oversea geologists as generally warm, but this must be accepted with reservations. At one time during the period the central portion of Australia sank so low that the sea came in from the north and divided it into two islands. Though coral fossils are absent from the rocks laid down in this sea, it is not thought nowadays, however, that it was a cold sea.

Nevertheless evidence from fossils and of other kinds shows there was climatic fluctuation during the Tertiary. About the middle of this period, particularly in the Miocene Epoch, the climate became warmer than today and remained so until the end of the period when the beginning of the Pleistocene Epoch foreshadowed the widespread refrigeration of the Glacial Age.

Throughout most of the Tertiary, the climate was moist, though fossil soils from that time show that the interior of the continent was already drier than the periphery. The "dead heart" was not, however, as extensive nor as arid as at present. During the Pleistocene there were times of cold climate when less of the rain falling on the land was evaporated and lost to the air. More water ran down the river courses, and lakes which are now dry formed in the far inland. These happenings may have been due to increased rainfall as well. Arid conditions intervened between these times of more effective or increased precipitation; the desert core of Australia extended, with rivers and lakes drying up and dunes developing. Over the last 50,000 years at least, there have been alternations of more and less effective precipitation.

Such drier and wetter phases have continued from the Glacial Age into Recent times.

It is interesting to look back and visualize what was happening to Australian plants during these climatic changes. It is from fossils buried at different times since isolation that we naturally look for information. Such information is unfortunately as yet very fragmentary. Fossils have been found in the deep leads, that is river alluvium buried beneath lava flows, leaves for the most part and fossil wood. Many of the leaves are comparable to those of the present rain forest, but their exact identification is very difficult. One German botanist many years ago with great enthusiasm classified them with living European genera such as oaks, elms, poplars and others, and advanced a theory of a universal Tertiary flora, but his views are now entirely discounted. Leaves of *Eucalyptus*, which are unmistakable, have been found in the deep leads, also *Melaleuca* and *Callistemon,* two common genera of shrubs today, showing that these typical Australian plants at least have a long lineage.

Much can be learnt from the comparative study of existing plants no less than of fossils. Apart from systematic classification into orders, families, genera and species, they may be divided into groups according to environment. There is for instance the rain forest, with plants of many different orders, but all with certain characters in common. The flora of the sand dunes shows another set of communal characters, as does that of bleak mountain tops, or that of the desert.

Apart from these communities there is in Australia, as elsewhere, what might be called the archaic group, plants descended from remote ancestors which were here before the arrival of the angiosperms. These are virtually living fossils and some have survived for ages with very little change. Among them are tree ferns and lowly plants of this type, the curious little Blue Mountain lycopod or mountain moss, cycads like *Macrozamia* and many pines.

If there be snobbery in the plant world the pines might well scorn other forest trees as mere parvenus. The Tasmanian beech might claim that its ancestors "came over with the conqueror", and eucalypts lay title to nearly as ancient a lineage, but the pines could aptly retort that they were here ages before either and that they were indeed once the aristocracy of all plants.

All pines are of ancient descent. There was a time prior to the coming of the angiosperms when they formed practically the whole forests of the world. They spread completely over all the land, and though they have been decimated by the onslaughts of more modern types, they still form large forests in North America, Europe and Asia. Elsewhere they are confined to small scattered communities, often isolated from each other by the whole width of continents and oceans.

In Australia each species of pine has its peculiar interest. Those in Tasmania are very different from those on the mainland. They are either peculiar to Tasmania, or, where relationship does exist, it is with other pines living in far distant localities. There is a small pendulous shrub called *Microstrobos*, a relative of the pines, living on the mountains near Lake St. Clair.

This is confined to Tasmania, but a closely allied species occurs in a small patch below the Leura Falls in the Blue Mountains in New South Wales. None are found elsewhere. Still more curious is a little pine, a straggly little shrub named *Diselma*. This lives only above 3,000 feet in the mountains of western Tasmania, and in the winter is often entirely buried in snow. Its nearest, and very close, relation is *Fitzroya*, which lives in the wilds of Patagonia. No similar pine is found in any intermediate locality.

The stately King William pine, found in the dense forests of the west coast of Tasmania, has no living counterpart in other regions, but it is almost identical with pines found fossil in the Jurassic rocks of Queensland, fossils over 135 million years old. The curious celery-top pine is also confined exclusively to Tasmania. It takes its name from its foliage, though what appear to be masses of small serrated leaves are not leaves at all but flattened stalks, to which the true leaves are attached as minute scales. The Huon pine is a noted timber tree living near the Huon and Gordon rivers and growing up to 135 feet in height. It belongs to the genus *Dacridium*, other species of which are found in New Zealand, New Caledonia, Borneo, Malaya and far-off Chile. Apparently all these pines once had a world-wide range, but have becomes generally extinct, and now only survive in a very few scattered localities.

The pines on the Australian mainland are no less interesting. The commonest are the cypress pines, of which there are many species, all but one living in the eastern States. These belong to the genus *Callitris,* which is confined to Australia, although two allied genera live elsewhere, one in north Africa, the other in south Africa. The botanist R. T. Baker has claimed that *Callitris* is the oldest of all the pines and that microscopic sections of the timber are practically identical with those of fossil wood from the Carboniferous rocks of Europe, rocks about 300 million years old.

Cypress pines form their own type of Australian bush, particularly on the western slopes of the tableland of eastern Australia. The white pine likes the flats. The black pine prefers the hills, the stonier and less fertile the better; it is indifferent if the rock be granite, sandstone or limestone. Here the black pine forms dense low forests or thickets, and the deep green of the foliage is conspicuous from afar in an otherwise grey landscape. Some of the pine forests are of large area, notably the Pilliga Scrub north of Coonabarabran, which is composed nearly entirely of black pine. Similar dense forests cover much of the Flinders Range in South Australia, where they push up right into the heart of the arid country. The white cypress pine also forms considerable forests in south-western New South Wales and Victoria, the larger trees being valued for their timber, which is resistant to the attacks of the white ant. Another species is found dispersed in the mallee scrub, and still others are found near the coast, one in the sandhills behind the beaches in southern Queensland.

Of the other Australian pines, the *Araucarias* also have an interesting geological history. There are two species, the hoop pine or colonial pine of the coastal regions in northern New South Wales and southern Queensland,

and the bunya bunya of Queensland. Both species form quite large forests. The bunya has extremely prickly leaves on the stems and branches and is a close relation of a South American species called the monkey puzzle, so prickly that "even a monkey would find it difficult to climb". The geographical distribution of the *Araucarias* is as peculiar as that of the other pines. They are found in New Caledonia as well as in Australia, in Chile, Brazil and Bolivia. The Australian species have apparently lived in their present location for long ages, for fossils very close to the bunya pine have been found in Jurassic rocks near Ipswich and in other neighbouring localities.

Two other pines must be mentioned, the so-called plum tree, *Podocarpus,* an inhabitant of the dense rain forests of the coast, and the gigantic Queensland kauri, also a member of the tropical jungle. The Queensland kauri is another link with New Caledonia and New Zealand, where closely related species form a large part of the main forests. The Queensland tree was never common and now seems doomed to extinction. It could never be overlooked, for individual trees were conspicuous by their huge size, and from any vantage point could be seen from miles away, towering like huge umbrellas above the level of the jungle top. I remember seeing one of these felled in the scrub near Tewantin. It was on a low hill-top, and its fall shattered the other trees in its path. From this tree five logs were cut, each 16 feet in length, the largest 11 feet, the smallest eight feet in diameter. The branches at the top were the size of ordinary trees, but were too bent to be worth cutting for timber at that time.

From the pines we pass to the angiosperms, the true flowering plants which form the bulk of the Australian bush. One of the most comprehensive papers on their development in Australia was written by the late E. C. Andrews as long ago as 1916.[1] This is a highly technical paper, and scientists are not unanimous in accepting all of Andrew's conclusions. Nevertheless his broad conception of what took place is probably within the bounds of reasonable accuracy.

Andrews considered that the first Australian flowering plants were all adapted to a warm damp climate and came from a tropical north. He says, "Australia appears to have been stocked with plants, both as luxuriant trees, as xerophytes,[2] and as dwarfed forms during the Upper Cretaceous. After the isolation from the world generally, many of the old luxuriant forms and the stunted forms were driven into the sandy wastes of extra-tropical Australia, and there, in their new surroundings, they developed numerous xerophytic species and even families, which from weak and modest beginnings gradually became hardy, vigorous and finally aggressive, and only limited to Australia by geographical isolation on the one hand, and inability to invade jungle areas on the other."

From this summary it is possible to visualize the course of events. It is a picture of grim and relentless war of the plants, a war that has gone on

[1] "The Geological History of the Australian Flowering Plants", *American Journal of Science*, 1916, Part iv.
[2] Literally, dry-living, botanically adapted to living in arid conditions.

ceaselessly for some 70 million years. There are two main armies opposed to each other, the massed battalions of the jungle and the guerilla forces of the xerophytes. When the climate was warm and moist the jungle advanced far and wide, but when prolonged droughts descended on the land the jungle was beaten back to its last fastnesses, to patches of basaltic soil on the tropical and semi-tropical parts of the east coast.

This is an extraordinary conflict, for not only do the main armies wage battle, but within their own ranks there is a ceaseless struggle for place. In the jungle there is continual striving to reach the sunlight above. There is also the fight between the larger plants and the innumerable vegetable parasites that prey upon them. The dark undergrowth is choked and overcrowded as each species seeks room to grow and expand. Yet if the tension were relaxed, if the individual striving ceased, the whole might easily cease to exist. The jungle crowds together for mutual protection.

Except right in the tropics, if too much light and air be admitted to the jungle, if spaces are opened up, most trees, if not all, are doomed to destruction. Many trees are lightly rooted in soft soil and bound together and supported by vines, and when this support fails and they fall they bring down many other plants with them. Much of the undergrowth is ill-fitted to withstand the direct rays of the sun, and when it withers and dies, the ground beneath becomes baked and hard, even in brief spells of dry weather. Shade and abundant moisture are minimum necessities for the jungle's existence.

A feature of the undergrowth in the Australian jungle is that a large proportion is composed of immature saplings of the surrounding species. Many of the seeds which fall to the ground from above germinate, but in the absence of sunlight soon wither and die. Others may survive for years in a state of deferred maturity waiting for the opportunity to reach the light above. Should a tree fall there is a sudden acceleration of growth and a race upwards to fill the gap. One fortunate sapling, perhaps more virile or better placed than the others, will win the race and expand its topmost branches to cut off the light from those beneath. In time it becomes another great tree, but the remainder of its generation are doomed to destruction unless another such opportunity should occur.

The Australian jungle or rain forest, with the exception of a few patches near Darwin and in Arnhem Land, is now confined to parts of the eastern coastal districts, and chiefly to areas where the soil is composed of decomposed basalt. There are really three zones of rain forest, the tropical jungles of Queensland and northern New South Wales, the warm temperate rain forests of the southern New South Wales and eastern Victorian coast, and the cool temperate rain forests of the southern Victorian highlands and western Tasmania.

The tropical jungle, according to Andrews, is of course the surviving portion of the original Cretaceous forest. In the long interval of time since then there have been some changes, new species and even genera having developed that are peculiar to Australia, but on the whole it still does not differ greatly from the jungle in other tropical lands throughout the world.

Many of the plants, if not of identical species, are at least closely allied to those of other lands. Above all there is a similarity in growth, and the jungle presents much the same appearance wherever it is found.

In eastern Australia the jungle is vernacularly called "the scrub", rather a misnomer, as scrub is more correctly applied to areas of small, stunted bushes. A somewhat more appropriate term used locally is "the brush". During the oscillating temperature and moisture conditions of the Pleistocene and Holocene, the jungle at times advanced from the continental margins into the interior and at others retreated towards the coast. The last change has been deciphered from the changing pollen rain which has settled on the bottoms of volcanic crater lakes in the Atherton Tableland. This fossil record has recently shown that in the period of 10,000 to 7,000 years ago the rain forest advanced westwards as climate grew warmer. However, when white men came, it was still restricted to a belt close to the eastern coast of the continent which had rich soil and adequate rainfall. Now much of it has disappeared or is fast disappearing before the axe and the plough. Unfortunately for itself, the jungle grows best on the richest and most fertile soil, and this land is the first to be cleared when civilization spreads. Great areas already have disappeared, on the Atherton Tableland in Queensland, on the coastal mountains and plains from Cairns to Townsville, the great Tewantin Scrub near Gympie, on the Dorrigo in New South Wales, and the Big Scrub which formerly filled the Big Bend of the Richmond River, about 70 miles across. Rich dairy farms and fields of sugar cane now take its place. Some is still left, particularly in scenic and faunal reserves, or on the sides of the mountains and in the deep and inaccessible gorges of the coastal rivers.

With the general passing of the scrub much that is picturesque and useful has disappeared, particularly in the way of valuable timber. The variety of the trees is very great, and the various species grow intermingled with each other, and are not segregated into communities or separate forests. This makes for uniformity over great distances. The trees grow close together, their foliage mingling 60 to 80 feet above the ground, so that looking over the forest from a hill-top all that is seen is a level surface of green. From the ground it is difficult to distinguish between the various species. Nearly all the larger trees spread out in wide, radiating buttresses. Their trunks are covered with the picturesque staghorn or elkhorn, with orchids of many kinds, and are festooned with tangled ropes of water-carrying vines. To the untrained eye the bark of many of the trees is similar, and their foliage is so densely mingled far above the ground that it is impossible to tell from which individual tree fruits and leaves have fallen.

There is very little undergrowth in the denser parts except young trees or trees with suspended growth waiting an opportunity to fight upwards towards the light. The air is damp and in perpetual shade, and fungi of innumerable varieties are everywhere, on the tree trunks, on fallen logs and on the ground. Fungi are the scavengers of the rain forest. They live on the decay of other plant life. When a tree falls, strangled by the parasitic figs, or from other causes, it is soon covered with a miniature forest of different fungi, and in a year or two is rotted away to add to the rich humus beneath.

It is in the slightly more open spaces that the undergrowth is thick and that progress is most difficult. Here the wiry lawyer vine spreads its tendrils, armed with thorns of needle-like sharpness to trap the unwary, and the broad innocent-looking leaves of the young stinging trees produce unexpected pain and blisters. Here also are the great green lilies or cunjevoi, the juice of which is proclaimed by the timber getters an antidote for the sting of the stinging tree.

One of the most remarkable trees in the jungle is the epiphytic fig. Commencing as a tiny seedling from a seed lodged in the forked branches of its host, it soon sends down a number of aerial roots to find further root in the soil below. These may form a veritable curtain or wall, or more commonly they may envelope the host in a network of confluent coils and ultimately strangle it to death.

Certain animal species are found only in the jungles of Australia although they may have relatives or actually occur in New Guinea too. Giant rodents, the tree-climbing kangaroos and that quaint marsupial the cuscus are confined to the extreme north of Queensland, but large tree-climbing frogs and the brown and green tree snakes come well to the south, as do many species of scrub pigeons, the brush turkey and other birds. The satin bower bird and lyre bird are also to be found, but this is not their only habitat, for they are found wherever the bush is thick. Land shells of many kinds are found in the rotting undergrowth, including the giant snail of the Richmond and Dorrigo, which is over three inches across. Not so attractive are the clouds of mosquitoes, the leeches that greedily await the passerby, and the ticks which insert themselves so insidiously into the skin.

It is unfortunate that with the passing of the jungle so much of Australia's timber wealth has disappeared. When the land was cleared timber was cheap, and distance and freight made it unprofitable to market. Thus millions and millions of superficial feet of magnificent building and furniture timber went up in flames. Red bean, black bean, silky oak, Queensland maple, crow's foot elm and a hundred other varieties, including even cedar, were thus sacrificed, never to be replaced.

Quite different from the tropical jungle, but also a true rain forest, is the forest which covers most of western Tasmania. Here the climate is cold, too cold for most of the jungle trees of the north, but with such a high rainfall that a very dense growth of forest is produced. This has developed its own peculiar plants, such as the *Richea* species which are members of the Australian heath family, Epacridaceae. One of them, restricted to Tasmania, is known locally as the grass tree and grows up to 40 feet, and others have a tropical palm-like appearance.

A connection with the original Cretaceous forest is of course the predominant Tasmanian beech, mentioned earlier in this chapter. No mention can be made of the Tasmanian rain forest without reference to that extraordinary plant, the horizontal or *Anodopetalum*. This plant alone has made the western Tasmanian bush almost impossible to penetrate. It is really one of the saxifrages, which are normally small alpine plants. It starts as a slender sapling in the space between the larger trees, and grows to a considerable height while yet only about an inch in diameter. It then falls to the ground with its own weight, branches grow upwards similar to the parent stem, these in turn fall and branch again, until the whole is an impenetrable tangle. In the mining districts beaten paths are made over and not through the horizontal, and it is possible to walk 10 or 12 feet above the ground. This can be a very dangerous proceeding, for if the traveller falls through a hole it closes above him, and without outside assistance he is hopelessly trapped. The bush rose, *Bauera,* with its dainty pink and white flowers, is not so treacherous, but its stems are very tough, and can also be so thickly matted as to be nearly impenetrable.

In the jungles and rain forests we thus see something of the original character of the Australian bush. In the next chapter the actual, the real, Australian bush will be dealt with, and it will be seen how it has developed particular features which make it unlike any other forest in the world.

The Bush Becomes Australian

Beyond the jungle flourish the xerophytes. This purely technical term, meaning literally dry living, in a botanical sense means plants which have developed means of conserving moisture. They are capable of withstanding either arid conditions as a whole or periods of drought, and of living on poor and porous soils. It is used frequently here, for it is both euphonious and suggestive. Its Greek origin conveys the idea of a handful of hardy Spartans opposing the luxury-loving hosts of Persia. We might speak of the unending struggle between these plants and the rain forest, with both sides contending for the domination of the earth, as the War of the Xerophytes.

The xerophytes have their battles among themselves. In their struggle for survival they must adapt themselves to hard conditions. They develop either tough and thick or long needle-like leaves, with stony cells and minute breathing pores, or thick bark full of insoluble kino or resin to protect the precious sap within. Their roots must be large and capable of penetrating deeply to reach the merest trace of moisture far below the surface. Fruits must be hard and woody and slowly maturing, seeds must be capable of retaining their fertility after lying for long periods on the sun-baked earth.

Even when rainfall is adequate, the xerophytes flourish on the poor porous soil of sand hills, on the infertile summits of sandstone tablelands, on the boulder-covered granite hills. It is from these that the main and unique part of the Australian flora has developed, the eucalypts, the wattles, the tea-trees, the casuarinas, the banksias and a host of others.

In the long period of its evolution this flora has adapted itself to many different environments, at the same time retaining its own peculiar facies. Over the whole continent there is great diversity both in soil and climate. Extremes of temperature are not so great as in some other lands, for there are no mountains above the line of perpetual snow, and no deep depressions like Death Valley in California or the rift valley of the Jordan River in Palestine. Parts of north Australia are well within the tropics, but the greater part of the continent, including Tasmania, is essentially temperate in climate.

The extreme range of humidity is much greater. Part of the east coast of Queensland has an annual rainfall of over 100 inches, as also has the western portion of Tasmania. The extreme north near Darwin and part of the Kimberleys has a moderate to a heavy rainfall, but it is concentrated into a brief wet season, and the remainder of the year is practically dry. The coastal regions of the east and south have a regular rainfall averaging from 30 to 40 inches. Farther inland the rainfall becomes less and less, and in a few areas there is practically none. In the far interior where the average is low, that is, down to a few inches a year, it is also irregular; it may be well above the average for a few years in succession, or there may be long periods of extreme drought.

In the whole of this vast region there are only a few limited areas of absolute desert which are entirely deficient in plant life. The total variety is astonishing. So far between 10,000 and 11,000 species of Australian flowering plants have been named, and even now, after 150 years of collecting, new species are occasionally being discovered.

Of this great variety by far the greater number of characteristic Australian species live in the poorest of soil. In Australia it may be said that the poorer the soil the more varied is the vegetation. It is on the sandy heaths and the sandstone hills near the coast that our great wealth of native flowers is to be found. Nearly all of these are peculiar to Australia. The neighbourhood of Sydney is in the heart of a sandstone area, and though the city has grown and the bush has been pushed back, it is still possible within a short radius to see more species of native flowers than are found in the whole of the British Isles. In Western Australia, within a short distance of Perth, the variety is still greater.

The disappearance of the bush before the growth of cities is perhaps inevitable, and bushfires started by human agency and vandalism have both played an ignoble part. The days have gone when from quite close to the city it was possible throughout the year and particularly in the late winter and spring to see the sandy heaths ablaze with an infinite variety of flowers. The days are gone also when from the road or railway on the Blue Mountains thousands of waratahs could be seen on either hand. Fortunately native plants are now protected by legislation, and it is an offence to pick them except on private land. Beyond the cities and a little off the beaten track they still flourish. On the deserted road from Nowra to Nerriga, a road cut off by an air field during the war, the open forest glows in the late winter with masses of boronia, with epacris, native fuchsia, grevillea, and innumerable other varieties, separated every few yards by the vivid red of the stately waratah. *Telopea* or "seen-from-afar" is the appropriate scientific name of this typical Australian flower.

The scent of the bush is surely a part of the scenery. It is said that Australian wild flowers have no scent, and while this is true up to a point, there are some which have a marked fragrance, such as the brown scented boronia of Victoria. The perfume of the green lily of the rain forest is very powerful, so strong in the close and shaded atmosphere as to be at times overpowering.

More subtle is the real scent of the bush, the mingled odour of minute quantities of aromatic oils given off by the leaves of many different plants. It is characteristic of xerophytic plants that while many of their flowers are scentless, their leaves and stems are particularly rich in essential oils. The ubiquitous eucalypt is of course prominent among these, and gum forests have a distinct characteristic odour. We are so used to this that it is apt to pass unnoticed, but for those who return to Australia from abroad it is at once a very breath of home. I remember when returning from an Antarctic expedition after more than a year in a land where there are no smells at all, our ship anchored overnight about half a mile from the shore at the south end of Tasmania. The air was still, and from across the water the scent of the gum trees was very distinct, so that even at night there was no doubt that we had at last arrived home.

Most of the essential oils are pungent but pleasant, a few such as the creeping boronia are rather unpleasant, and one or two of the acacias are actually fetid. In contrast is an eriostemon which grows on the barren sandstone hills near Denman in New South Wales. The leaves of this graceful plant have a very high yield of oil almost identical in odour with attar of roses.

Eucalyptol may be considered the characteristic odour of the bush. In spite of its medicinal associations, it is really pleasant, particularly when it is just a suggestion in the air and blended with the scent of the other bush plants. Eucalyptol is usually associated with the gum tree, but not all members of the genus *Eucalyptus* contain this particular constituent. In fact the chemistry of the group is complex and varied, each species having its own particular essential oil which is constant in chemical composition. Some, such as the bloodwoods, have little oil at all, others are rich in peppermint, and are known in the vernacular as "peppermints" or "peppermint gums". One at least, the Queensland "lemon-scented gum", has a very strong lemon flavour and odour, hence its scientific name *Eucalyptus citriodora.*

Of the many trees, shrubs, creepers and lesser plants composing the bush it is impossible within the limits of one chapter to mention even briefly more than a few. In selecting what are the most characteristic groups, the most important is undoubtedly the order of the myrtles, or botanically, the Myrtaceae. To this the eucalypts and tea-trees belong. The next place might well be filled by the Proteaceae, an order nearly confined to Australia but for which there is no popular name. Belonging to the Proteaceae are such well-known plants as banksias, waratahs and hakeas or needle bushes. The third place is given here to the wattles, part of the world-wide genus *Acacia*, itself a member of the legumes or bean-producing plants. Then come the Casuarinaceae, commonly called oaks (which they certainly are not), she oaks, forest oaks and swamp oaks. Though most species occur in Australia, the family is also found in south-east Asia and is represented on some islands in the Indian Ocean. Then there are such plants as boronias, Christmas bells, flannel flowers and hundreds of others, some belonging to families and orders with a universal distribution, but which have developed their own peculiar Australian characteristics. Mention must also be made of the desert plants, the spinifex and saltbush of the far interior.

All these plants are xerophytes, part of the great army which has waged successful war on the rain forests and now dominates Australia. Many of them are descended directly from luxury-loving plants, plants needing rich soil and abundant moisture, but they have developed into a hardy virile race, well able to resist encroachment. Some have even turned aggressors and in turn invaded the margins of the jungle, where they hold their own in the struggle for existence.

The order of the myrtles is a splendid example of how a group of plants originally belonging to the jungles has for the most part turned xerophytic, and gone over, as it were, to the ranks of the enemy. Its present distribution is peculiar, for while a limited number of species may be found in many tropic lands, with only one in southern Europe, the great bulk of living species is confined to two areas, tropical America and Australia. The tropical species are, however, very different from most of the Australian in that they possess a soft, pulpy fruit while the fruit of Australian species is generally a hard woody capsule. An exception in Australia is the lily-pilly (*Eugenia*) which possesses soft fruit, and as might be expected all species are confined to tropical or warm temperate rain forest.

Amongst the Australian myrtles the eucalypts are of course pre-eminent, and it is probable that 75 per cent of all the trees in Australia are "gum trees" of one kind or another. The *Angophoras,* known vernacularly as "apples", are so close to the eucalypts that they may be considered with them. The tea-trees, a loose term, which may be widened to include the paper-bark or *Melaleuca, Callistemon* or bottle-brush, *Leptospermum, Baeckia* and other genera, also have a very wide distribution in Australia, and are peculiar to it. Though rarely attaining the size of large trees, they play almost as important a part as the eucalypts in the bush. One myrtaceous tree with a woody fruit varies somewhat in its habitat from the others. This is a species of *Tristania,* the brush box, which has successfully invaded the jungle.

The genus *Eucalyptus* is by nature practically confined to Australia. There are about three hundred different species in the continent, rather more than one half living only in the east. A few are common to the east and west, but the bulk of Western Australian species are not found beyond the borders of that State. There are some species in New Guinea, an odd one or two in Malaya, and curiously enough one indigenous to the Philippines.

Through man's agency eucalypts are fast becoming acclimatized through-out the world. Their value as hardwood and their extremely rapid growth make them particularly suitable for reafforestation, and others such as the Western Australian red flowering gum, are greatly favoured as ornamental trees. In California, forests of eucalypts have become so well established that many Americans are beginning to look upon them as their own. When the American troops were here during the war, more than one expressed surprise at the way they had spread in Australia, and would not be convinced that this was their original home.

The species range in size from the giant gum and mountain ash of Victoria and the karri of Western Australia, which are amongst the tallest trees in the world, towering sometimes to over 300 feet, to the small curious many-stemmed mallees of the mallee scrub. Others are small stunted trees living on the highest mountain tops in the snow country in Tasmania, in Gippsland and in the Monaro. Others live as tall straight trees in the moist valleys of the coastal country, sometimes in dense forests on the flats, or elsewhere finding root in the most precarious of situations on almost precipitous rocky slopes.

On poor sandstone country and on the high plateaux trees are generally smaller and of irregular growth. On the western slopes forests are of medium size and rarely as dense as in the coastal areas, and on the western plains and in the far interior trees become fewer and fewer. Even in the semi-desert there are here and there belts of eucalypts, rarely very tall, but often thick trunked and with wide-spreading branches and thick foliage, lining the generally dry watercourses. These trees recall the famous lines in what is almost our national song:

> Once a jolly swagman camped by a billabong,
> Under the shade of a coolabah tree.

The coolabah epitomizes the spirit of the outback. It is a eucalypt. *Eucalyptus microtheca,* to give it its scientific name, a tall tree up to 80 feet in height, with a rough ashy grey bark, found only in the far west, from Narrabri through western Queensland and into the Northern Territory.

Andrews considered that the eucaypts originally developed from a soft-fruited and soft-leaved myrtaceous plant, and slowly adapted themselves to poor soil, hot sun and drought. R. T. Baker,[3] on the other hand, held that the bloodwoods and angophoras are the most primitive types and are survivals of the original stock. The question is unsettled, but it is probable that Andrews is nearer to the true solution. There has been much difference of opinion among botanists upon other points, particularly upon the constancy and validity of the various species.

Having had the opportunity for many years of testing it in the field, my own view is that Baker's work is easily the best for the amateur botanist, and by it the various species are easily recognized in the field. Among other things the theory was advanced that eucalypts may be classified by the chemistry of their essential oil, and that there is a definite relationship between the nature of the oil and the veining of the leaves.

If a gum leaf is held up to the light, small round clear patches may be seen. These are the oil glands, many in some species, few in others. The leaf may vary in shape, but there is always a strong midrib, and from this secondary veins branch on either side. In the angophoras and bloodwoods these veins are almost at right angles to the midrib and there are few oil glands. In such trees as the Tasmanian blue gum the veins are at an angle of about 45 degrees, there are many oil glands, and the oil is rich in the well-known

[3] R. T. Baker and H. G. Smith, *Research on the Eucalypts and Their Essential Oils*, Second Edition, 1920, Technical Education Series No. 24, Government Printer of New South Wales.

eucalyptol. In the peppermints the secondary veins are almost parallel to the midrib, the oil is abundant and has a strong peppermint odour. There are of course many variations and intermediate stages, and an account of the complicated chemistry would be highly technical and out of place here. That there is a relationship between the chemistry of the oil and the veining of the leaf may, I think, be accepted.

An interesting fact pointed out by Baker is the geographical division of the eucalypts according to the colour of the timber. It is certainly curious that the light-coloured timbers are nearly confined to the south and the red timbers to the north, with an intermediate belt in which the colours overlap. Baker considered that the red-timbered species are the oldest and that the white timbers later developed from them. This theory is open to considerable doubt.

Another interesting feature of the eucalypt is the leaf stalk or petiole, which is generally twisted. This results in the partial rotation of the leaves which keeps the hot sun on their edges, probably reducing transpiration and the loss of valuable moisture.

The juvenile or "sucker" leaves provide further material for thought. In most species the leaves of young plants as well as those growing from broken stumps and branches and from trunks charred by bushfires are quite distinct from those of mature and normal trees. They are generally soft and glaucous, often heart-shaped and stem-clasping, more suggestive of the leaves of the jungle plants than those of the xerophytes. In both animals and plants ancestral characters are often preserved during embryonic and juvenile stages, and the juvenile leaves may well give a clue to the ancestry of the eucalypts.

A few species of eucalypts actually retain the juvenile type of foliage throughout their whole life, and never develop what is considered the normal adult leaf. One of these is the Argyle apple (*Eucalyptus cinerea*). The thick glaucous foliage of this picturesque tree is a feature of the open forest on the tableland between Marulan and Goulburn in New South Wales. Another small species living only on the highest mountain tops in Tasmania does the same.

Fossil evidence on the eucalypts is very meagre. In the absence of fruits and buds, odd leaves when found give very little clue to the species to which they once belonged. Specimens of such leaves personally collected from beneath the basalt at Armidale had the intermediate venation suggesting a eucalyptol oil, but more than this could not be determined. The exact age of these fossil beds is unknown. but they are probably mid-Tertiary, and if so, the eucalyptol-bearing group is at least of considerable antiquity.

The eucalypt is at its greatest magnificence in southern Australia, in Tasmania, in Gippsland and in the south-western corner of Western Australia. Here the giants of the forest are to be found. They are amongst the tallest trees in the world, comparing in height if not in girth with the famous Sequoias of California. There have been reports of dead trees which had been 400 feet in height, and records of three over 300 feet are well authenticated. In Western Australia, at the little settlement of Manjimup, there is a living karri 286 feet in height, and in the neighbouring forest the karris average 250 feet, and are up to 10 feet in diameter, their smooth, white trunks rising with but little taper to 200 feet before the first branch is reached.

Within this corner of Western Australia, between Perth and Albany, are some of the finest forests in Australia. Areas are still virgin, except where the tracks from the sawmills push their way into the shady depths. A few yards from the beaten track are the primitive wilds, just as they were before the coming of man. Care must be taken when away from the track, for visibility is restricted, there are no landmarks, and any but a good bushman will soon lose his sense of direction. The air is still and sound is hushed as in the cloisters of a cathedral. The chirping of birds is distant and far overhead. On every hand the great white columns rise vertically from a bracken-covered floor, the only undergrowth young trees struggling upwards towards the light. Here is one phase of the real Australian bush.

In Gippsland far to the east are other great forests. From such a vantage point at Mt Tinghiringhi on the New South Wales-Victorian border, the view extends southwards for 50 miles to a distant horizon. This is rugged terrain, though from the height of 5,000 feet the countless gorges are dwarfed to bands of deeper green. In all this expanse no trace is visible of any village, human habitation or road. Such as do exist are well hidden, sunken in the sea of vegetation. This also is the place of great trees; it is the home of the giant mountain ash (*E. regnans*), of the alpine ash (*E. delegatensis*), and of the brown stringybark (*E. obliqua*) and many others.

Across Bass Strait, in Tasmania, some of these species form tall forests here also. The most accessible is that on the slopes of Mt Wellington, in full sight of the capital of Hobart nestling at the mountain's base. The big trees here live at an elevation of over 2,000 feet. They are not quite as large as in other parts of the island, but many are over eight feet in diameter and well over 200 feet high. Dwarfed at the feet of the big trees, but still of their own kind gigantic, are the tree ferns, often 20 feet and even 30 feet in height. Higher on Mt Wellington, from the Springs at 2,500 feet, other species of eucalypt appear, species typical of the mountains of the rugged interior and west coast. Here they are beautifully zoned, each species within a well-defined belt, the trees becoming smaller and smaller as the mountain is ascended, until those living on the summit above 4,000 feet are little more than shrubs. These are the mountain gums (*E. coccifera*), conspicuous by their thick milky blue leaves.

The forests of great eucalypts continue right up the east coast of Australia, filling the plains, climbing up the sides of the gullies, covering the summits of the tablelands and spilling over the divide into the interior. The species change from district to district and the trees tend to become smaller away from the coast. On sandstone and granite and on the exposed tops of mountains they become small and stunted, while the boxes and ironbarks on the western slopes often have a widely spreading and drooping habit. In the dense coastal forests the trees are tall and straight, and here are many of the most valuable timbers, tallowwood, and so-called mahogany, spotted gum, blackbutt, blue gum, grey gum, ironbarks and many others. Though not as large as the southern giant gums and mountain ash, many are huge trees. I have seen tallowwoods 12 feet in diameter, and trees of six to eight feet are quite common.

Forests of eucalypts continue far out into the western plains. In the interior, belts of the stately Murray River red gum, water gum, coolabahs and a few other species mainly line the banks of the rivers or follow the dry watercourses, which only run after occasional rains. Even in the dead heart of the continent there are few areas without a patch or two of gum trees to remind the rare wayfarer that he is still in Australia.

Before leaving our national tree mention must be made of a peculiar and characteristic type of Australian bush, the mallee scrub. The mallee is an inhabitant of the near outback. To the city dweller it is but a name, to the early explorers it represented nearly waterless country difficult to penetrate, to the farmer it is country which when cleared is suitable for valuable, drought-resisting wheats. Alas! clearing has often proved in the long run disastrous. In the mallee, soil exposed by the removal of its protective vegetation has been particularly vulnerable to wind erosion. In dry seasons northerly and westerly winds have blown away many feet of the surface, sometimes right down to the subsoil. In the red dust-storms millions upon millions of tons of soil have been carried away far over the eastern and southern States and even out to sea. Dust from this source has been recorded from as far as New Zealand, borne by the wind across the Tasman Sea.

The mallee nevertheless still covers thousands of square miles in Western Australia, in north-western Victoria, south-western New South Wales and South Australia. It proves again the adaptability of the eucalypt to varying conditions. For the mallee is a eucalypt, or rather a number of eucalypts, since there are many species, red or water mallee, bull mallee, blue mallee, narrow-leaved mallee, and others which are just mallees.

All have one feature in common. Instead of rising in a trunk direct from the ground, there is a large underground rootstock from which many branches rise to form a spreading bush. This may be from 10 to 20 feet high. Individual clusters may be separate so that it is possible to pick a way between them, or they may be so close and interlocked that they are practically impenetrable.

The mallees are inhabitants of the plains, preferring a sandy or gravel soil in that belt of country having an annual rainfall of from 10 to 15 inches. One natural home is about the lower reaches of the Lachlan, Murrumbidgee and Murray rivers, and in Victoria they cease at the margin of the rich basaltic western plains. They are true hardy Australians, despising the amenities of rich soil and abundant rainfall.

In the mallee scrub are many other plants, not eucalypts, but all Australians of hardy lineage. There is one of the cypress pines (*Callitris propinqua*), various melaleucas and tea-trees, and a curious little plant, *Exocarpus* or broom mallee, which forms leafless bushes about six feet high. Another plant, *Loudonia,* though it has no popular name, is a beautiful sight in October and November, when its green stems are capped with golden-winged woody fruits. There is war in the mallee scrub as elsewhere in the vegetable kingdom. Even here are parasitic plants, and the clinging stems of the creeper *Cassytha* are often so thick as to quite smother their hosts.

The mallees, though too small for timber, have their distinctive economic uses. Some of them are very rich in eucalyptus oil. In the Wyalong district the blue mallee has been extensively used for oil distillation. The best yield does not come from the old trees; so these are rolled and crushed by heavy rollers and then fired. In from 12 to 18 months there is a thick new growth, and from this a much greater yield of valuable oil is obtained. A popular local use for the mallee is as firewood, for the rootstock once alight does not go out until it is entirely consumed, leaving only a fine white ash.

The mallees have a certain interesting geological association, for much of the mallee country coincides with the area covered by an incursion of the sea into southern Australia in Miocene times, that is, about 25 million years ago. When the land sank this sea extended beyond the present junction of the Murray and the Murrumbidgee rivers. The bedrocks underlying much of the mallee scrub were deposited in this sea. From the cliffs along the lower Murray River and from bores put down in western Victoria, extinct sea shells, teeth of sharks, the ear bones of whales and many other fossils have been obtained. However the mallee does not usually grow directly on these marine rocks but on lime-rich sands deposited during drier Pleistocene and Recent stages in dunes running from west to east.

Second only to the eucalypts in giving character to the Australian bush is the great genus *Acacia,* familiarly known since the early days of settlement as the wattle. In the more densely settled parts of the eastern States, the golden bloom of the wattle is familiar to all, but it is in the outback that the acacias have their greatest significance. In the language of the bush, the mulga or the myall conveys a picture of illimitable distances, of far horizons, of dusty cattle and of drovers estimating the distance of one waterhole from another. Yet myall and mulga are not only types of country; they are the native names of species of *Acacia,* names which have definitely been added to the Australian vocabulary. Far beyond the limits of cattle, on the fringe of the desert, in the desert itself, right across the continent, mulga is to be found, a dense bush with grey green foliage. On the margins of human habitation, stock find nourishment in its tough leaves, and the native make boomerangs from its hard dark wood. Millions of acres of the interior are covered by dense mulga scrub, and it is truly a real part of the Australian bush.

As with mulga the brigalow is not only a type of country but a species of acacia (*A. harpophylla*). There are dense forests of brigalow on the downs country of central Queensland, and the brigalow scrub is as well known in the north as the mallee in the south. Brigalow is not of great height, but another acacia of the western plains, the ironwood (*A. excelsa*), is a tall tree, 50 feet or more in height, and is noted for its beautiful hard red timber.

There is one acacia of the interior, the gidgee (*A. cambagei*), also known as the stink wattle, whose timber gives off a most fetid odour. In pleasant contrast is the Western Australian acacia known as raspberry wattle, whose wood has a strong odour of raspberries. Amongst the largest acacias is blackwood, famed for its beautiful timber, ranging from Tasmania to Queensland, and up to 70 feet high. In further contrast are the small prickly acacias, growing on the coastal sandy heaths or on the dry tops of sandstone tablelands.

These are but a few of the great number of species. There are of course the well-known wattles, prized not only for their flowers, but also for the tanning properties of their bark. Such trees as *Acacia decurrens,* the green or black wattle of New South Wales, and *Acacia pycnantha,* the golden wattle of South Australia, are but two of the many, whose full description would be more appropriate in a book on Australian botany.There are even more species of acacias than of eucalypts, and in the world nearly eight hundred are known, of which about five hundred are indigenous to Australia. Of these five hundred about four hundred belong to one of the divisions of the genus, the *Phyllodiae,* which are confined to Australia and make our wattles different from those elsewhere.

Even to the non-botanist the division of the acacias into two main groups by their foliage is at once apparent. There are those such as the green wattle, which have soft drooping feathery pinnate leaves, and there is the main group with really no leaves at all but phyllodes, flattened stems which serve the functions of leaves, generally lanceolate in shape, but sometimes reduced almost to thorns. This feature is characteristic of the group, but some interesting evidence of their possible origin is afforded by the foliage of the seedlings and immature plants, which is almost invariably of the soft, pinnate type. If this embryonic character has been inherited from remote ancestors, then it is probable that the acacias with phyllodes are the most recent and peculiarly Australian development of the genus, and that those with pinnate leaves are surviving descendants of the ancestral stock which lived in the original Cretaceous forest.

Andrews in a lengthy paper on the origin of the leguminous plants holds this view,[4] and he considers from this and other evidence that the acacias originally came from the north, and that "the ancestral forms could be conceived as trees and shrubs of luxuriant type, flourishing in moist warm climates, and all possessed of pinnate leaves".

It is unfortunate that space does not permit some account of the many other interesting groups of plants which inhabit such special areas as our sandy heaths, the gully bottoms, the alpine meadows on the higher mountains and the plains of the interior; but before leaving this chapter brief reference must be made to that interesting Australian tree, the *Casuarina.*

The casuarinas are best developed in Australia, and though by no means a large element in the bush, they are conspicuous in their own special habitats. Large areas of swampy country near the mouths of rivers and tidal flats just above the mangrove zone are occupied by forests of the swamp oak

[4] *Journal of the Royal Society of New South Wales,* 1914.

(*Casuarina glauca*), while several species are found both dwarfed and as trees in the sandstone country. Most conspicuous of all are the large trees of the river oak (*C. cunninghamiana*), up to 100 feet high, lining the banks of rivers on the coastal plains and the western slopes. There are 29 species of *Casuarina* in Australia, 13 of them confined to Western Australia, where their habit is similar to that of the eastern species.

The casuarinas are true xerophytes, and like so many other plants they have developed peculiarly under Australian conditions. They have practically dispensed with leaves, and the thick foliage consists of slender drooping branches, longitudinally ribbed and jointed at short intervals, the true leaves being reduced to rings of minute scales at the joints.

With the casuarinas we leave the Australian bush, leaving its greater variety still undescribed. Sufficient has been said, nevertheless, partially to answer the question asked at the beginning of the previous chapter, that is, whence did it come and when? We see that certain elements such as the pines have been with us for countless ages, that the bulk of the true flowering plants came from adjoining land in the Cretaceous, and that after land connections with other countries ceased at the end of the same period the plants of Australia developed in isolation into virile hardy types, many of unique form. Now with the coming of man a new flora has been introduced, some of it in the form of noxious weeds. It is pleasing to say that in the struggle for existence the indigenous flora has so far been well able to hold its own.

Ancient Lands of the West

To the early explorers Western Australia seemed a harsh and austere land. The approach nowhere offered an attractive façade to suggest a land of promise beyond. After a brief glimpse of an arid coastline the early Dutch navigators gave its reef-strewn waters a wide berth, as did Matthew Flinders who followed them. Even when the fertile plains and forests of the south were discovered long afterwards the first settlers found it a hard land to woo and win. The potential was always there, but it has taken over a hundred years for patient pioneers to lay the foundations of a modern civilization. The day of the pioneer is not yet over, and even now the hardy, the adventurous and the ambitious are pushing farther and farther from the closer settlements into the Kimberleys or into the far interior, to the verge of the Great Sandy Desert itself. Little by little nature is being harnessed to the service of man, and in the process there has been built up a virile and a kindly race.

Western Australia is not only the largest but undoubtedly the oldest of the Australian State divisions. From north to south it covers 20 degrees of latitude, from well within the tropics to the temperate regions facing the Southern Ocean, yet scenically it is practically a single unit. That is not to say that it lacks either interest or charm, yet if a surplus of detail be omitted it can be dealt with in the limit of one chapter. Its topography, though lacking in height, is on a grand scale, its units so large that it needs detachment from immediate surroundings to appreciate them. The Western Australian is attached to his environment as much as any other Australian and with as much reason. He is nevertheless mainly an inhabitant of a narrow coastal belt, and that in the extreme south-west of the State. Four-fifths of the vast area of the interior has been seen by few, parts of it by none. Until recently there were still large areas of unexplored desert, great expanses of sand dunes that had been seen only from the air. Explorers, prospectors and odd stockmen had ventured at times far into the wilderness, and some never returned. However in recent years mapping of the rocks by government geologists, search for valuable minerals including oil, and the operation of great rocket ranges from Woomera have resulted in men penetrating practically everywhere with the help of four-wheel-drive vehicles and helicopters.

The newcomer who travels by rail or road for any distance will be astonished by the absence of many features to which he is normally accustomed. There are no high mountains, and the wide tableland covering most of the State rarely rises to any great height. There are no large rivers, and many of the short coastal streams are intermittent in their flow. There are no lakes, unless the many salt pans of the interior, generally dry, be described as such. Except in the south-west there are no large forests, and most of the country is covered with straggly eucalypts, salt bush, spinifex, mulga and other drought-resistant plants. The plains are vast and seemingly endless, relieved only occasionally by residuals of the hardest rocks.

One of the main clues to Western Australian topography is quite close to Perth. Perth is a garden city on a noble estuary set in the midst of a fertile plain. Looking across the city from King's Park an unbroken line of high ground can be seen a few miles to the east and extending to the north and south as far as the eye can see.

This is the so-called Darling Range, which is not a range at all in the sense that a range of hills should have a downward slope upon the other side. It is rather an escarpment, the sharp edge of a tableland which stretches continuously nearly half-way across Australia. If the edge of the escarpment is traced on a map it will be seen as an almost straight line running north and south. In the south it cuts across the base of the peculiar hammer-shaped peninsula which has as its northern and southern extremities Cape Naturaliste and Cape Leeuwin; it passes just to the east of Perth, and ends many hundreds of miles to the north about 50 miles inland from Geraldton. It is never very high, averaging about 1,000 feet, sometimes precipitous, sometimes sloping more gently, and often broken by ravines where the coastal streams come down to the plain.

This line is the dividing line, the line between the new and the old. On the western side is a world of recent change. At some time in the past the surface of the tableland extended much farther to the west than it does now. Then its margin gave way and it sank down, forming now a fault, now a gentle fold. In a later age the eastern side of the tableland probably rose somewhat along the same line of weakness or faulting. When the coastal belt sank it carried with it the Collie coal seams which had been deposited in swamps in the Permian Period and other formations which now lie deeply buried beneath Perth and other centres on the coastal plain. This narrow strip of coastal country has since been subject to further change, chiefly the formation of the coastal limestone described later in the chapter on coastal scenery.

Beyond the summit of the Darling Range a region is entered where time for countless ages has stood still. This is the land of Yilgarnia, Yilgarnia the timeless, the mysterious, a land of such antiquity that figures alone are beyond the conceivable. Yilgarnia is of the oldest lands of the world. It was here not only long before the advent of man, but before there were birds or reptiles or fish or even the humblest of animals, before there were plants, indeed as far as is known before there was any living thing but for the very simplest of forms made up of a single or a few cells.

The land is still here, though its face changed many times before it assumed its present appearance. When masses of rock were first squeezed upwards from the bed of the primordial ocean high mountains no doubt rose into the air, mountains of stark rock, desolate and dead, the only movement

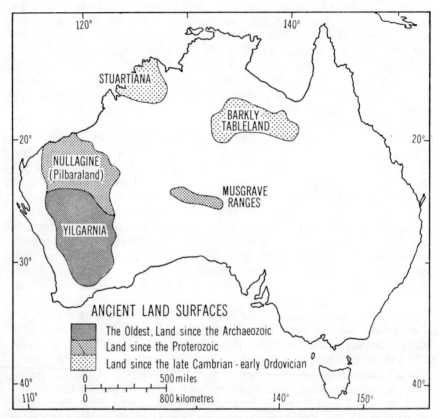

2.   Map of approximate extent of the most ancient land surfaces of Australia.

the rushing of rivers or avalanches of snow and ice from the heights. In the course of time these original mountains were worn away, and there were other changes, probably the submergence of the land beneath the sea on more than one occasion. Of what took place through the aeons which followed there is the merest glimpse, for nearly all is hidden behind the veil of incredible antiquity. Even when all the changes had merged into a stable and general landscape it was to our conception still in the dawn of geological time.

In the nineteenth century geologists were content to commence the geological record or sequence of periods with the Cambrian Period, that is the oldest period where there is definite evidence of living things in the sea. All events prior to the Cambrian Period were simply called Pre-Cambrian, and it has only been in recent years that an approximate sequence has been worked out for the still older formations. Pre-Cambrian now means a succession of periods, each of vast duration, grouped into two vast eras, the Archaeozoic and the Proterozoic, each representing a lapse of time greater than from the Cambrian to the present day.

It was in the earlier part of the Archaeozoic Era that this land of Yilgarnia was born, and it was throughout the Archaeozoic that changes occurred which may now be partially traced. After the close of the Archaeozoic it looms into a stable and permanent shape, a land surface amongst the oldest in the world. In bare figures we may say that it came into existence about 2,600 million years ago, suffered various vicissitudes for about 1,500 million years, and remained virtually unchanged from then till now.

It is but a fragment of the old land which still remains intact. To the west and south much has sunk beneath the sea, while to the north it has sunk and been buried beneath later formations. The Recherche Archipelago off the southern coast consists of part of the submerged land which is now disappearing before the erosive power of the sea. What was left still remains in practically the same condition as it was during the Proterozoic Era. Its surface has crumbled into soil or been thinly covered and protected by wind-blown sand, while a few residuals of harder rock are here and there exposed. It has in later ages been clothed by countless races of plants. Subjected again and again to changing climates, it has been ragged and threadbare in times of aridity, but arrayed again in luscious green when rain fell and rivers ran.

What is left of Yilgarnia occupies an area roughly rectangular in shape, 400 miles from west to east and about 700 miles from south to north. The transcontinental railway, after climbing the Darling Range, enters upon a vast plain, rising slowly to an elevation of about 1,500 feet, then descending as slowly to the surface of the Nullarbor Plain.

Eastward the nature of the country is reflected in an ever-diminishing rainfall. The coastal rivers rise near the western fringe of the tableland in a settled, afforested country with a moderate but adequate rainfall. Even when the permanent rivers are left behind the forests continue far into the arid interior. Chief among the trees is the tall and beautiful salmon gum (*Eucalyptus salmonophloia*), its bright salmon-coloured bark conspicuous in an oft-times drab landscape. The salmon gum covers much of the country about Kalgoorlie, where the rainfall is about 12 inches, and even farther to the east it forms a belt right on the edge of the Nullarbor Plain, marking, as it were, a boundary to the ancient land.

North and east the country is more arid still, and the eucalypt forest gives place to the hardier plants of the deserts which fill the eastern part of Western Australia. The surface forms a plain with an area of nearly 300,000 square miles. A few elevations bear the names of mountains, a comparative title only, for most are mere hills a few hundred feet above the general level.

Beneath much of this is a solid bastion of ancient rocks. Since a slight elevation of the whole region, probably towards the end of the Tertiary Period, the level has been reduced by some hundreds of feet, and the few hills are the residuals of the hardest of the rocks. The country varies slightly with changes in the old geological formations. The commonest rock is granite or gneiss, which on the surface has decomposed readily, the quartz sand left over covering the plain so that there are few rocky outcrops. Where the granite does outcrop, it is in low rounded masses from a few yards to several miles across. Dykes of quartz porphyry which penetrate the granite are very resistant and are weathered into rugged hills covered with sharp and angular fragments. Schists like the granites have decomposed to some depth and rarely outcrop on the surface. Another coarsely crystalline rock, black in colour, akin to granite in structure but of different composition, is gabbro. This occurs as occasional intrusions, which, being very hard and resistant, are left as hills.

Recent geological mapping has shown that over much of the easternmost part of Western Australia towards the state border there are varying thicknesses of Permian sedimentary rocks on top of the ancient crystalline basement. These are partly glacial deposits laid down on land by great glaciers at this time and partly marine deposits of cold seas into which icebergs brought some glacial morainic materials. Much later in Cretaceous times the sea formed a great strait from the Eighty Mile Beach to the Nullarbor Plain and deposits from it survive over large areas of eastern Western Australia.

The scenery in the north and east offers little attraction, unless it be the wide open vistas, the clear air and the freedom from urban conventions. The country might well have remained in its solitude if it had not been for the lure of mineral wealth hidden within the rocks. The story of the golden cities of Coolgardie and Kalgoorlie is outside the scope of this book, except that it emphasizes the part the prospector has played in taming the wilderness. Civilization owes a debt to the men who led the way into the unknown, facing the hardships of hunger and thirst and the peril of a lingering death. No less a debt is owed to the enterprise and engineering skill that brought water by pipeline 300 miles across country to the new city and gave its people security from thirst.

Gold is not the only precious substance found in the area. A feature of the scenery is the great number of salt pans, some of them of very large extent. Salt pans are frequently figured on maps as lakes, but they are merely shallow depressions, sometimes containing a little water after occasional heavy rains. The depressions are formed by the irregularity of wind-blown deposits of sand, for wind, unlike water, does not deposit sand in a perfectly horizontal bed. Hollows are formed, and after the rare intervals of heavy rain they may remain for a time as considerable lakes. Since evaporation is very rapid they soon dry, and as they have no outlet, any salts in the water remain behind, and may accumulate into extensive beds. Not all the salt pans so common in this part of the continent are due solely to the irregularity of wind erosion (also termed deflation) and deposition. Many of them, especially the larger ones, have a more complicated history. Lakes such as Lakes

Weathered sandstone. Near Kununurra, East Kimberley, Western Australia. *J. Cavanagh, CSIRO Land Research.*

Waterfall on granite in Sir John Forrest National Park, Darling Range, Western Australia. *Western Australian Tourist Development Authority.*

Stirling Ranges and surrounding xerophytic scrub, Western Australia.

Lefroy and Cowan, each over 50 miles long near Coolgardie, are arranged in chains along broad shallow valleys inset in the plateau. These represent ancient drainage lines to the south coast, which have been stopped functioning in this way by drier climate when the rivers ceased to flow along them with sufficient regularity and force to prevent the wind dropping sand across their paths to act as natural barrages.

The deposits of salt generally consist of common salt, gypsum, and occasionally the rarer potassium salts which have considerable commercial value. Such a deposit is in the bed of Lake Campion, a salt pan on the fringe of the wheat belt about 170 miles east of Perth. Here it is estimated that there are 10 million cubic yards of a clayey material containing 63 per cent of alunite, a complex compound of potash and alumina. There is sufficient potash here to supply the whole needs of Australia for over a hundred years, as well as providing oxide of alumina as a by-product and potential source of aluminium. Nearby in Lake Brown, gypsum is already being worked on a scale which makes it one of the three noted points of supply for Western Australia.

In the south the ancient lands show more variety in their topography where they fringe upon the Southern Ocean. Unlike the western edge there is no sharp escarpment to divide the ancient from the newer world. When the southern edge of Yilgarnia subsided—probably in the early Tertiary Period —it sank gently over a wide area and formed a gradual slope down to and below the sea. Towards the end of the Tertiary the sea bed rose again above sea level, and bore on its surface some 200 feet of marine limestones and sandstones which had been deposited in the interval. Inland from Albany is a wide plain sloping gently to the south, and beneath this are the marine rocks deposited in the Tertiary sea. These rocks extend in a belt along the coast to near Ravensthorpe, nearly 200 miles to the east.

At Albany, between the plain and the sea and extending into the sea itself, is a line of hills and rocky headlands. These consist of red granite and gneiss and are really outlying fragments of the old land of Yilgarnia to the north. Albany itself lies in a hollow between two rounded hills of gneissic granite over 600 feet high, facing the placid waters of Princess Royal Harbour, with the wider waters of King George Sound to the east. Everywhere along the adjacent coast are these hills of gneiss and granite, more or less rounded by time, but descending to the coast in precipitous cliffs. The waves have undercut the cliffs into rugged and grotesque shapes, carving out caves and grottoes into which the breakers burst with a resounding roar. Great stacks or towers of rock stand isolated from the shore, and numerous rocky islets and hidden reefs bestrew the inshore waters.

Some of the headlands are of considerable height. Mt Gardner, on the peninsula forming the eastern tip of King George Sound, is 1,308 feet high, Bald Island farther to the east and at the extreme end of the rocky belt rises over 1,000 feet, while between is Mt. Manypeaks, 1,843 feet. King George Sound is nowhere more than 20 fathoms in depth, and like so many Australian inlets owes its existence to the rising of the sea after the end of the Great Ice Age.

The origin of the high belt of ancient rocks at Albany affords many problems, as does that of two still higher ranges beyond the plain to the north. For the most part this plain is perfectly flat, covered with low timber and scrub, or by large areas of swamp, with here and there shallow depressions and salt lakes. Through it all the Kalgan River meanders slowly to the sea, emerging first into the nearly enclosed waters of Oyster Harbour and thence into King George Sound.

Northward across the plain and about 20 miles from Albany is the spectacular Porongurup Range. The range is small, only about eight miles long and two miles wide, but is very conspicuous as it rises like an east-west wall over 2,000 feet directly from the plain. Composed of granite, it is remarkable for the regular way joint planes have split the rock into huge blocks and slabs. Some of these slabs have sheer walls hundreds of feet high and are so regular and sharp-edged that it would seem some giant sculptor had shaped them with hammer and chisel.

From the Porongurup Range a magnificent view is had northwards across 20 miles of plain to the irregular outline of the Stirling Range, which completely fills the horizon from east to west. In a region where mountains are low, the much greater height of the Stirling Range makes it difficult to fit into the general topography. Like the Porongurup Range it runs east and west; it is sharply separated from the plain to both north and south, but it is of much greater extent, 40 miles long and from three to four miles wide. From the distance it appears as a chain of higher peaks separated by lower saddles, fading away to the west, but higher in the centre and higher still in the extreme east, where it terminates in the highest peaks of all. Of these Bluff Knoll is 3,597 feet high, and Ellen Peak, right at the termination of the range, about 3,420 feet. From close at hand the mountains are seen to be very dissected and worn, and there are several gaps a little above plain level.

To the north of the range the scrub-covered plain continues at a higher elevation, averaging about 800 feet, always sloping gradually upwards and merging into the ancient land surface of Yilgarnia. It is a desolate region, with innumerable small salt lakes and but little habitation. The climate is arid, and the vegetation that of the intermediate zone between the great coastal forests and the real desert plains of the interior.

This is the home of one of the most curious Australian plants. From the distance the eye is caught by what seems to be the hostile figure of a native with spear erect. Then another is seen, and another, until the whole bush is filled with warriors, immobile but alert and watchful. A nearer view dispels the illusion, and they are seen to be nothing more than grass trees or *Xanthorrhoea* plants, with the appropriate popular name of blackboys. The resemblance to a native from even a short distance is startling. The rough trunk, blackened by bushfires, is about the height and girth of a man and is capped by a cluster of grass-like leaves which resemble a thatch of unkempt hair. The flowers are borne on a stalk rising from the summit of the trunk, and simulate a spear. *Xanthorrhoea* is closely related to the lily family and there are seventeen species in Australia, several of them confined to the west.

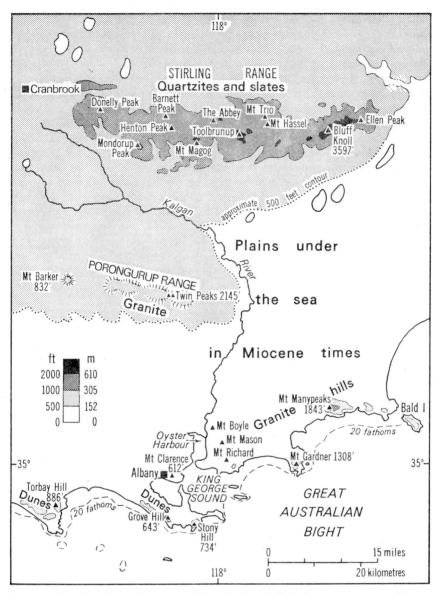

3. Map showing alternating belts of high land and plain from the coast at Albany to the Stirling Range.

The Stirling Range is not composed of granite like the other high land in the area, but of sedimentary rocks of somewhat later age, though still within the remote Proterozoic Era. The rocks are thin beds of hard quartzite, separated by red shales and slates. They are not greatly folded, and lie where they were originally deposited on the worn surface of the granite beneath. It is interesting to note that there is another range composed of ancient sedimentary rocks about 100 miles east of Albany. This is the Mt Barren Range which stretches from Doubtful Island Bay to Mary Ann Harbour. It is of considerable height and narrow, runs east and west, and though lower than the Stirling Range, is almost in an exact line. The rocks here are schists and also overlie the granite.

The origin of the Stirling Range and other high land in the Albany district has long been the subject of discussion. The simplest explanation is that the hills and mountains are the more resistant remnants of a former plateau in the area which has been subject to much erosion. However this does not account for the regularity of three parallel ridges, all running west to east with uniformly low land between, nor the particularly great height of the Stirling Range above the plateau to the north. Residuals of a dissected plateau tend to be rather uniform in height and scattered about irregularly.

Other factors are therefore involved. In the case of the Stirling Range, it has been shown that this relief feature is a horst, a solid block of country pushed upwards and bounded by great west to east thrust faults (Fig. 4). The thrusting must have taken place very far back in the past and the present relief is the result of later differential erosion, not the effect of the original displacement. Whether faulting is involved in the case of the Porong-urup Range is not known. Sometimes it is wider spaced jointing, which causes certain granite areas to stand out as ridges from more closely shattered rock around, otherwise rather similar.

The high lands are now again being worn away. For a time they were probably islands in the Tertiary sea. Those still on the sea coast are being eroded both by the sea and from above: those inland are being lowered by

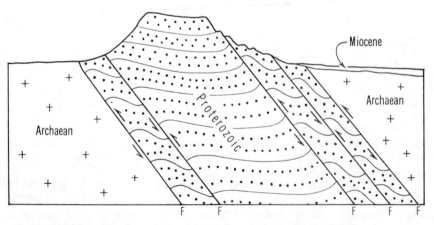

4. Geological section through the Stirling Range (after R. T. Prider).

frost, rain and the natural decomposition of the rocks themselves. The mountains now probably show but a remnant of their former grandeur.

There is another old land surface in Western Australia, not so ancient as Yilgarnia, but nevertheless so old that it has remained as land from before the beginnings of historical geology. Much of what happened in the early days of the world's history is hidden in the mists of antiquity, but here and there the veil is drawn aside and it is possible to glimpse strange landscapes, and even reconstruct some phases of the changing geography. Thus it can be stated that prior to the Proterozoic Era, Yilgarnia was an island covering south-western Australia, and that another ancient island now known as Stuartiana lay to the north, over much of what is now the Kimberleys.

Between these two islands in Proterozoic times was a sea, a sea which persisted intermittently through many long ages. There were many changes in its geography: sometimes it was deep, sometimes shallow, sometimes its bed was upheaved and contorted, even temporarily converted into dry land. Many great rock formations were deposited in this sea, the last of them the Nullagine Series, a great thickness of conglomerates, quartzites and calcareous shales which still cap the tableland. These rocks are themselves incredibly old, and when they were finally elevated above the sea, it was still long before the Cambrian Period and the beginning of the story of living things upon the earth.

The northern shore of Yilgarnia lies about the present 25th parallel of south latitude, much of it in an uninhabited and partly unexplored region. To the north of it is this other old land composed of the rocks deposited in the Proterozoic sea. It is now a plateau, part of the main plateau of Western Australia, but lifted to a somewhat greater altitude.

This part of the tableland lies approximately between the 21st and 25th parallels of latitude, and reaches from the narrow coastal plain far into the interior, where it flattens out somewhat towards the Great Sandy Desert. It is an uneven tableland, broken up into blocks of varying height by faulting, and further dissected into many gorges and deep valleys by erosion. The Fortescue, Ashburton, Minilya and Gascoyne Rivers, with lesser streams, still plunge downwards from the tableland to the coast, but they are intermittent, and for months and even years their beds are dry. There is evidence that in the past, probably in phases of more effective precipitation in the Pleistocene Period, they were mighty and permanent streams, and it was then that they carved out the deep gorges in which they flow.

There is much interesting scenery on the summit of the tableland, which has been lowered by some hundreds of feet since its elevation. Most of the higher points are remnants of the hardest rocks and variation in their shape is often due to difference in the bedding of the sedimentary rocks which compose them. Mt Bruce, the highest mountain in Western Australia, 4,024 feet above sea level, is according to J. T. Jutson a residual left over by erosion. One interesting mountain is Mt Augustus, 3,627 feet, situated near the head of the Lyons River, a tributary of the Gascoyne. This mountain is a huge block about five miles long from south-east to north-west and two miles

wide, standing some hundreds of feet above the general level and composed
of very hard gritty conglomerates. The strata at some time had been folded
into a sharp arched fold or anticline, steeper on its northern side, and the
shape of the mountain now closely follows the bending of the strata, as shown
in Figure 5.

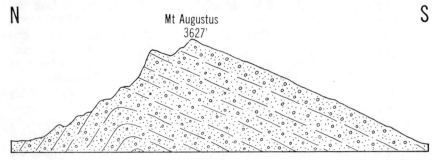

5.   Section of Mount Augustus showing how the shape of the mountain reflects the
folding of the rocks (after Maitland).

Here again Jutson is of the opinion that the mountain is a resdiual of erosion,
and not just folded above the surrounding country.[5]

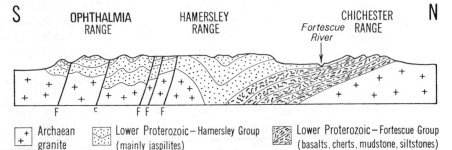

6.   Section through the Ophthalmia, Hamersley and Chichester Ranges showing
different types of scarps. The Chichester Range is part of the Nullagine Tableland.

The most northerly part of Pilbaraland as this area of Australia was called
by Jutson consists of the Hamersley Range and the Chichester Range or
Nullagine Plateau separated by the broad valley of the Fortescue River (see
Figure 6). Jutson thought this valley was a rift valley between two horsts.
However both the Fortescue valley and the escarpment on the northern side
of the Nullagine plateau, where it opens onto the upland plains of the Desert
Basin, are due solely to erosion. There are faultline scarps in the area as
shown south of the Hamersleys in Figure 6.

All this region until recently was remote and rarely visited, though there
have been large sheep stations in the lower coastal belt for a long time. In

[5] J. T. Jutson, *The Physiography of Western Australia*, Bulletin No. 95, Geological
Survey of Western Australia.

the uplands it has again been the lure of mineral wealth that has brought habitation into the wilderness. Farther to the north, at Marble Bar, gold, tin and other metals were the magnets; in the Hamersleys it was the famous blue asbestos which first led to permanent settlement.

There are many scenic gems in the mountains which the aeroplane has brought within the tourist's reach. The summit of the tablelands is flat and covered with a straggly scrub of spinifex, with here and there the black and white trunks of the curious eucalypt, black heart, or the rougher-barked bloodwoods. From above, the gorges open unexpectedly like great gashes, vivid and vibrating with colour. Their walls are everywhere composed of perfectly horizontal strata, lying undisturbed in the same position as they were when deposited almost at the beginning of things. They are ledged and terraced where harder bands have resisted erosion, but sometimes they rise sheer from the creek bed to the rim of the tableland above. The rocks though not folded have been chemically altered by time, and are a riot of colour, reds and blues and mauves. In contrast is the olive green of cool pools in the depths. Sometimes far below may be heard the tinkle of running water. Here the white trunks of slender river gums struggle upwards towards the light, and between them deep green patches of delicate maiden-hair ferns are clustered upon the rocky ledges.

The blue asbestos was recently the economic mainstay of the district. The main centres of production were the picturesque Yampire and Wittenoom gorges, which enter the main valley of the Fortescue River from the south, and wind for many miles with almost vertical walls deep into the heart of the tableland.

Asbestos is a curious mineral, a complex silicate in composition, crystal-lizing into a closely packed mass of long flexible fibres. It is here derived from the chemical alteration of the sedimentary rocks themselves, and is interbedded with them in seams up to three inches across. There are different levels where the asbestos is richest, sometimes high on the face of the cliffs, sometimes just above the valley floor. Mining was at first haphazard, but later the industry was systematized with adequate capital, tunnels were driven deep into the hills and a well-planned village at Wittenoom Gorge provided comfort and amenities to the mining population.

Asbestos has now practically ceased to be produced and has paled into insignificance beside the tremendous economic development which has fol-lowed the investigation of huge iron deposits in the area. The Nullagine series of Proterozoic rocks include many chemically precipitated sediments here, including some jaspilite formations rich in iron. This is now worked in open cuts at Mt Goldsworthy, Mt Tom Price and Mt Newman, to which the word colossal can be applied without any exaggeration. This has led to much port, road and railway construction, to fresh settlements as well as rapid growth of older centres such as Port Hedland.

We now leave our story of the ancient lands of the extreme west, but in the next chapter it will be continued with an account of other ancient lands in the far north.

Ancient Lands of the North

In March 1883 a party of men set out from Tuena in New South Wales with two wagons drawn by bullock teams, 50 horses and 700 mixed cattle. Their objective was good land that had been reported in the far north-west of Australia by the Forrest brothers on their Kimberley expedition. The full story of this epic journey has been adequately told by others,[6] and it was certainly one of the greatest droving feats of all time. It is a story of unfailing courage in the face of drought, hunger, thirst and many encounters with hostile natives. In the wilds of Arnhem Land the three MacDonald brothers were all that remained of the original party. One by one the others had become discouraged and returned to more civilized parts. Near the Roper River Charlie MacDonald became so ill that it was necessary for Donald to take him on to Darwin and thence back to New South Wales. William was left to go on alone, and he accomplished the amazing feat of reaching the Kimberleys with the surviving stock. He arrived and settled at Fossil Downs with 370 cattle and a few horses, after a journey lasting three years and six months.

This was nearly 70 years ago, but Kimberley is still a remote region and to the average Australian little more than a name. In many ways it is the Cinderella of Australia. There are still descendants of the MacDonalds at Fossil Downs, and there are widely scattered cattle stations in the belt of lowlands surrounding the central mountain core, but its economic wealth is still largely undeveloped, and its scenic beauties are unknown to the general public.

It is not difficult to prophesy a wonderful future for this part of Australia, when its rivers will be dammed, when irrigation will bring the fertile soil to fruition, and when communications will be improved so that the products of the soil can be transported to the centres of population. For the moment it is still a land of promise. Nevertheless, among the scattered inhabitants there are many who have unbounded faith in its destiny, and who strive hopefully for the day when the tide of fortune must surely turn.

Such is Kimberley, or rather the Kimberleys, for there is an East and a West Kimberley, and the plural is generally used for what is really one geographical unit clearly distinguishable from the rest of Australia. Most

[6] Mary Durack, "Kimberley Epic", Australian Geographical Magazine, *Walkabout*, February, 1948.

Ord River dam site in gorge cut through the folded Carr Boyd Range, East Kimberley, Western Australia. *J. Cavanagh, CSIRO Land Research.*

East Kimberley Plateau and the Ord River crossing at Ivanhoe, Western Australia. *J. Cavanagh, CSIRO Land Research.*

Devil's Marbles near Barrow Creek, Northern Territory. Corestones weathered from granite. *CSIRO Land Research.*

of it is a knot of rugged if not very high mountains projecting in a wide bulge into the Timor Sea. The coast is deeply indented, numerous gulfs penetrating far into the land, gulfs in which the rise and fall of the tides is 20 feet and more. Surrounding the mountains is a belt of lowlands wherein is much of the country's fertility. Through the lowlands the Ord and the Fitzroy rivers encircle the mountains in two embracing arms.

To the south and south-west of the fertile belt is the desert country, wide plains becoming more and more arid as they stretch into the interior, to merge at last into the Great Sandy Desert. Sandy deserts stretch almost unbroken from the Eighty Mile Beach to the Nullarbor Plain; south of the Great Sandy Desert comes the Gibson Desert, named by Ernest Giles after a companion who was lost there, and further south still the Great Victoria Desert, dedicated by Giles to the ruling monarch of his time.

From east to west the Kimberleys are 500 miles wide and from north to south 300 miles. A good deal of the centre of this great area is still unexplored. In one way the region is an anomaly, for with an apparently adequate rainfall the great problem is still that of water supply. The annual rainfall is from 20 to 50 inches, but it nearly all falls in the four summer months when the north-west monsoon blows. For the remaining eight months rain very rarely falls, and the rivers shrink from mighty floods to a mere trickle, or cease to run altogether. The sky is then clear and cloudless, the rich soil of the plains is turned to dust beneath the relentless sun, and on the river flats droves of wallabies compete with the cattle for the scanty feed. The pressing problem of the Kimberleys is the conservation of the copious summer rains by the construction of dams in the gorges, and the irrigation of rich soil which will in the future help to provide for the needs of a hungry world.

A very modest beginning was made some years ago at Liveringa on the lower Fitzroy River but this has been overshadowed by the more substantial start made to use the waters of the Ord River in East Kimberley. Despite many difficulties faced by settlers already using waters from the diversion dam to grow cotton, work is proceeding with a big dam on the Ord to irrigate a very large area of alluvial land on its lowest reaches.

The seasonal nature of the rainfall is reflected in the vegetation and also in the general scenery. Though well within the tropics there are no rain forests as in the same latitude in northern Queensland. The low-lying country and the plains are nearly treeless, covered with grass such as Mitchell and Flinders grass, or the porcupine grass known locally as spinifex (*Triodia*). This is quite different from the grass of coastal sand dunes, botanically true *Spinifex*. Stunted eucalypts sparsely cover the mountains, strange eucalypts which shed their leaves in the dry season to conserve moisture. Here also are many acacias and other xerophytic plants, all adapted to resist long periods of drought. It is only in the deep gullies and along the river beds that the vegetation is rich, and here the cajaput, one of the paper-barks (*Melaleuca*), is conspicuous with its graceful drooping branches and its rich soft green leaves.

On maps and for purposes of administration the Kimberleys are divided into an East and a West Kimberley, but scenically a more natural division is into the central mountainous core and the surrounding lowland belt. The division is not only of scenery but of time itself. The central highlands are the remnants of an old old land; the lowlands are the result of changes which took place in much later but still very ancient periods.

In the mountains of the Kimberleys it is difficult to trace the many changes in topography which took place in the very earliest periods, that is before it finally emerged as dry land, never again to be submerged beneath the surface of the sea. All the rocks within the area are classed as Pre-Cambrian; that is, they were formed in some of the many eras of the vast space of time before the beginning of geological history proper. There are granites which are probably the equivalent of those forming the ancient land of Yilgarnia near Coolgardie, there are altered sedimentary rocks comparable with those of the Hamersley Ranges, and there are gneisses and schists of periods probably intermediate in age between these two. There are also remnants of gigantic lava-flows which are still below the lowest of the Cambrian beds.

Here then is another of those old land surfaces which have survived for incredible ages. What its original appearance was we do not know. It was almost certainly a dead land, for while there may have been life in the sea, there is evidence that it did not appear on the land for long ages after.

Here and there the veil is lifted just enough to allow a glimpse of what was happening, and curiously enough, one of the things of which there is evidence is climate. It is known that at one stage at least the climate was intensely cold and not hot as surmised by many physicists for the early history of the world. Among the Pre-Cambrian rocks in Australia as in other parts of the world are beds of boulder clay and glacial erratics to tell of the first known of the great ice ages through which the world has passed.

It is known also that long prior to the Cambrian Period this part of Australia was an island, separated by a wide sea both from the ancient land of Yilgarnia in the south and from other lands to the east. This was the island of Stuartiana, so-called after Stuart the explorer, a man who did much to reveal the geographical secrets of the central deserts. Again, before the Cambrian Period the bed of this sea was elevated to land once more and Western Australia was united into one land. In the Cambrian Period itself, the sea again invaded the land and this time divided the whole of Australia from north to south, but the Kimberleys were not separated from the main mass of Western Australian for a very long period. In early Permian time a deep gulf penetrated deeply between the Kimberleys and the south-western part of the state which included Yilgarnia and Pilbaraland. Then in a great transgression in Cretaceous times the sea reached right from the Eighty Mile Beach to the Great Australian Bight to divide these two parts completely.

Parts of the ancient lands of the Kimberleys are visible from afar, particularly from the east. From the southern portion of the Ord River, the edge of the central tableland can be seen extending in a nearly vertical wall of uniform height as far as the eye can reach. This wall is of quartzite of Pre-Cambrian age. To the south the great Antrim Plateau is also visible from the upper Ord. The plateau here is compact and but little dissected, and capped by sheets of basalt which flowed over the country before the beginning of the Cambrian Period.

Nearer to the centre of the tableland the country varies; sometimes the summits of the highland are flat and even, while elsewhere it is deeply dissected by the numerous rivers which have their source there. Many ranges appear on the topographical maps, the Princess May, the Synott, the Artesian, the Durack and the Phillips ranges. According to Jutson these are not true ranges at all, but residuals of the main tableland left between the deeply excavated river valleys. Almost in the centre is Mt Hann, 2,547 feet above sea level. Its flat-topped summit stands about 800 feet above the general level, and from the neighbouring flat grass-covered country it is a conspicuous landmark. Its origin is not certain, but it is thought to be a residual of erosion, a fragment left over from the time when the tableland was even higher than it is now. The highest point is Mt Ord with a height of 3,070 feet in the King Leopold Range, the quartzite rim of the southern margin of the Kimberleys.

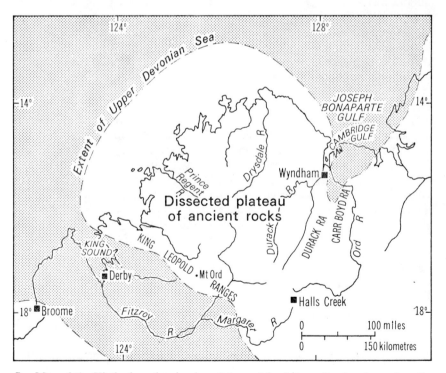

7. Map of the Kimberleys showing how it has retained its peninsular shape since the Devonian Period.

After the formation of the Pre-Cambrian land surfaces there is a long interval of which there is no record, but no doubt the face of the land changed considerably as the ages passed. The next picture of the topography is in the Devonian Period, some 380 million years ago.

In the opening part of this chapter Fossil Downs was mentioned as the locality where the MacDonalds settled after their epic journey across Australia. The name stirs the imagination, and indeed emphasizes how the Devonian Period has left its mark on the present topography. In the bed of the Fitzroy River at Fossil Downs are slabs of limestone composed entirely of the remains of extinct corals and other organisms which lived in the Devonian seas. Similar limestones occur along the southern margin of the Kimberleys, sometimes rising for some hundreds of feet above the general level of the plain. Such are the Napier Range north of the Fitzroy River and the Lawford and Emanuel Ranges to the east of it. In these ranges monoliths of limestone are often the solid remains of one-time Devonian coral reefs.

On the east side of the Kimberleys the Ord River cuts a picturesque gorge through the Pre-Cambrian rocks of the Carr Boyd Range; to the north nearer the sea there are Devonian reef limestones east of the river. These rocks give a fairly clear picture of the Kimberleys in Devonian times. The general outline was not greatly dissimilar from that of the present day. We can see it as a high mountainous peninsula protruding from the north of Australia. Both to the east and the west the land had sunk, and the Devonian Sea had invaded the land, a warm tropical sea, its shores fringed by extensive coral reefs. The sequence at one stage is very clear, and Dr C. Teichert has worked out how the positions of the reefs changed as the land sank, and how a barrier reef remote from the shore was eventually smothered by the deposition of sand.

The mountains in Devonian times were probably high, but their exact appearance can never be known. It is probable that they were bare and stark, with cliffs of red granite and black schist, capped with ancient and massive black lavas. They were also probably still lifeless, for land plants had not been long in existence. At the most, rocks at lower levels may have been covered with moss and lichens, and there may have been some ferns and quaint club mosses living in the gullies and coastal swamps.

After the Devonian Period the sea bed rose again, and in the long interval between the Devonian and Permian periods, the land took approximately its present shape. Then the bordering land sank again, and the sea penetrated deeply, particularly on the western and southern sides. The area of the present Desert Basin was deeply submerged, and at the same time the Kimberley Peninsula was again elevated, and the worn relics of the Devonian mountains subjected to a further cycle of erosion.

The new Permian mountains were doubtless less forbidding than those which had gone before; there would have been forests of primitive pines upon their slopes and dense mats of fern-like plants upon the plains and in the swamps. The climate too was different. This was another of the world's great ice ages, and huge ice sheets formed over many parts of the southern hemisphere. Evidence of glaciation has been found in Antarctica, southern Australia, South Africa and South America, and as far north as the Kimberleys and even India. No such evidence has been found elsewhere in the northern hemisphere. The study of paleomagnetism has helped to explain this. Various rocks rich in iron have their own magnetism which they acquired when they were formed. For example, rocks solidifying from a molten state acquire magnetic properties in accordance with the earth's magnetic field at the time of solidification. If these properties are measured, they enable the former magnetic field and the position of the magnetic poles at that time to be reconstructed, and from them the geographical poles. Thus it is known that the Permian South Pole lay at only 45°S in the Great Australian Bight and this explains the distribution of climate indicated by the Permian rocks of Australia.

Glaciers descended from the Permian mountains of the Kimberleys right into the sea. Here they broke away as icebergs, eventually melting and depositing their loads of rocky debris on the ocean bottom. Great boulders of granite and other rocks carried by ice from the Kimberleys may still be seen embedded in the neighbouring Permian marine rocks, and with them the remains of shells and other sea animals which lived at the time. Left behind on the land also were thick beds of glacial tillite and boulder clay which remained after the ice finally disappeared.

The Permian Ice Age again wore away much of the high land of the Kimberleys, and during long subsequent ages the land lay at a comparatively low level, so low that it remained free from further erosion. The surrounding land which had been under the Permian Sea rose so that it was slightly above sea level, and this portion of Australia assumed very nearly its present shape.

During the Tertiary the Kimberleys were progressively warped up above the level of the surrounding country to about 3,000 feet. To this uplift and to the extensive erosion which has since taken place the present topography is due, its details varying according to the different hardness and the complex geological structure of the various ancient formations forming the high lands.

Although there are great faults separating the Kimberleys from the Desert or Canning Basin to the south and from the Joseph Bonaparte Gulf Basin to the east, these are ancient lineaments along which movements took place many times no doubt but not recently in geological history. The various escarpments around the Kimberleys are erosional features due to varying resistance of the rocks. Limestone ranges such as the Napier and Oscar Ranges have indeed quite cliff-like margins due to solutional undercutting of this rock. Overall the Kimberley Tableland appears to be dome-shaped, sloping from the centre outwards to the marginal scarps. The numerous rivers rise near each other in the centre and radiate outwards towards the lower land. Many, such as the Margaret and the Leopold, flow southwards in deep gorges to join the Fitzroy, which sweeps westwards in a great arc round the southern base of the tableland before turning slightly to the north-west to enter King Sound.

The deeply indented coastline of the tableland and straight valleys running into its heart such as that of the Prince Regent River are undoubtedly influenced by parallel faults in a north-west to south-east direction. Again these features are not young landforms due to recent earth movements but the indirect product of older faulting, weakening particular zones and bringing rocks of different degrees of resistance side by side so guiding later erosion. The archipelago of islands which everywhere fringes this coast and its many gulfs are due to the drowning of the rugged, fault-guided relief of this old land of tough rocks by the general rise of the sea at the end of the Pleistocene Period. The archipelago represents the hills of harder rock left by erosion, which alone surmounted the higher sea level. Offshore the sea floor of the continental shelf, the Sahul Shelf, is unusually deep around coral reefs such as Scott and Seringapatam Reefs; downwarping is indicated by this which may also have affected the coastlands as well, though the Kimberley core is thought to be rising to balance this.

The rivers which flow south into the Fitzroy have cut deep narrow gorges through the great escarpment at the edge of the tableland. The gorge of the Margaret River keeps almost a straight line as it cuts across the part of the escarpment known as the King Leopold Range. This gorge has already become scenically famous. The King Leopold Range itself is a massive bastion of tough Proterozoic quartzitic sandstone overlying much older crystalline rocks. The gorge of the Lennard River through the Napier Range is also very picturesque. The walls of this winding Winjina Gorge (or Devil's Pass as it has also been called) are vertical cliffs over 200 feet high, dropping right into the river in parts. Much of the rock is corroded into vertical channels or chimneys. Above the gorge the plateau is flat though intricately carved in detail. The rocks are bedded Devonian limestones which reveal the interior structure of an ancient coral reef.

The upper portion of the Fitzroy itself has cut a spectacular gorge through the Geikie Range. Here also the rock is Devonian limestone in part horizontally bedded, and the nearly vertical cliffs have been worn into some extraordinary forms. In some places the bedding planes of the massive limestone have divided the cliffs into a series of steep terraces, one above the other, with bushes and small trees clinging precariously wherever they can find a hold. The solvent action of rain has been the chief agent in shaping the rocks. As it has run down the face of the cliffs it has everywhere produced a regular fluting, at the same time rounding the terraces above. Viewed from a little distance the effect is of a wide series of cascades and waterfalls, though for eight months in the year the rocks are quite dry. The edges of the cliffs above are rugged and castellated, corroded into innumerable picturesque and grotesque forms.

On the eastern side of the tableland conditions are somewhat similar though there is much less limestone. The Ord River is at times a mighty stream, and as it flows north is fed by many tributaries coming from the high tableland to the west. It rises in comparatively low land, but as it nears the sea it passes through much rugged country. It has indeed cut a gorge through the Carr Boyd Range which lies athwart its course. The explanation of this has not yet been worked out and more than one cause is possible. However it may be that this range rose across the pre-existing river course but so slowly that it was able to maintain its original path instead of being diverted far to the east.

Near here the tableland is not of great elevation, and much of it has already been worn to base level. It consists of plains or near plains, covered with low scrub and spinifex, with rugged remnants of Pre-Cambrian lava-flows and Cambrian limestone and of Devonian sandstones and limestones. These form numerous low hills and ranges. Typical of these is the Burt Range, also Mt Napier and Mt Panton, just across the border in the Northern Territory.

From the Kimberleys let us travel some distance to the east and seek still another ancient land surface, a land not nearly so old, yet venerable enough to rank among those few portions of the earth's crust which have survived unchanged since the dawn of historical geology. This is the Barkly Table-land, a region of special interest. Though some hundreds of feet above sea level it is a tableland by courtesy only. There are none of the features commonly associated with tableland topography, no high flanking escarpments, no deep gorges, no swiftly flowing rivers. It is indeed almost a perfect plain, practically free from surface rocks, with a deep uniform grey-brown soil, covered with Mitchell grass and with very few trees. To the south the plain slopes imperceptibly until it merges into the vast flat interior. The Georgina River and its tributaries rise from an almost imperceptible divide and meander sluggishly to the south. Often they are little more than chains of waterholes, and except in occasional flood times they fade away completely before they reach the South Australian border.

To the east the country is a little more rugged. Here the Barkly Tableland is bounded by a series of rocky ridges which run generally in a north and south direction. This narrow belt of country runs from eastern Arnhem Land in an arc southwards to well beyond Cloncurry. It is in itself a region of tremendous interest. The rocks are very ancient, comparable to those in the the great Kimberleys in age, and within them are great mineral deposits worked in the mines of Cloncurry and Mt Isa. This high belt of country has through the ages been, as it were, a barrier between the east and the west. During the Cambrian Period it formed the shore of a sea which lay to the west; in much later times it was again a shore for prolonged periods but then the sea lay to the east.

On its northern border the Barkly Tableland falls away rather more steeply to the coastal plain. Here its tableland character is a little more apparent. There is a broken escarpment two or three hundred feet high, and the waters of the Gregory River have cut gullies and miniature gorges into the underlying limestone. Though not on a grand scale the scenery is spectacular. Deep pools are enclosed at the feet of overhanging cliffs, there are rugged bluffs, and dense tropical vegetation chokes the hollows.

This is a sparsely settled country, with few roads and no railways, given over to large cattle stations. The one town, Camooweal, lies just about the centre, within the western Queensland border. Its height above sea level is 788 feet, which is about the average height of the plateau. In all its expanse there are no prominent hills nor valleys to provide landmarks.

Yet it is this absence of feature which provides the clue to its greatest interest. In most regions an absence of hills and valleys is due to the wear and tear of ages. The forces of erosion have ground down the mountains until nothing remains but their worn stumps. Here, in contrast, is a region where, extraordinary to relate, there has never been a mountain since before the Cambrian Period. There has never been hill or valley. The country has remained a plain period after period, era after era, while elsewhere whole lands have been engulfed by the sea, and the bed of the ocean has been elevated into alpine ranges, not once, but time and time again. Only for a brief period in the Early Cretaceous did the sea transgress the northern parts of the Barkly Tableland and the Mt Isa ranges. Remnants of marine deposits from that time reveal this, together with level summits in the ranges. Yet the inner parts of the Tableland escaped even this short-lived marine incursion.

It is a wonderful thing to contemplate these fragments of archaic lands and to separate them from the ever-changing landscapes which surround them. Here in the Barkly Tableland geological history ceased virtually at its dawn. The story is confined to one period. In this, the Cambrian, about 600 million years ago, the land sank and Australia was divided into two islands. The Cambrian sea was shallow, and in it accumulated a series of shales, dolomites and limestones. Then the sea bed rose again to form dry land. In central Australia, South Australia, Victoria and Tasmania, the rocks deposited in the Cambrian sea have since been subjected to tremendous stresses; they have been elevated into mountains, worn away again, sometimes again submerged beneath the sea. On the Barkly Tableland, however, the rise was gentle and but slightly above sea level, and then, save for a further slight rise in a much later period, all movement ceased. Never has the land been high enough to permit erosion on any scale, and beneath a protecting layer of soil the original rocks lie undisturbed just as they were originally deposited.

Everywhere below the surface are rocks of Cambrian age, shales and dolomites for the most part, and all so fresh and unaltered that they resemble in appearance the rocks of the very latest periods. They are revealed in a few rare outcrops on the surface, but mainly have been penetrated by wells and bores on the scattered cattle stations. In the rocks are beautifully preserved remains of the sea creatures which lived in Cambrian times. These are amongst the very oldest fossils known in the world, and they afford a glimpse of what this early life was like, even if they do not show its actual beginnings. In view of its age it is surprising how complex this early life was. All the creatures are of course extinct, but many can be correlated with things now living, though belonging to different orders and families. There were curious little crustaceans called trilobites which crawled upon the sea bottom, there were sea shells and the probable ancestral forms of starfishes and sea urchins;

but such highly organized animals as fishes were not to appear for many ages to come.

Though the original surface of this ancient land is still intact, it is possible that more recent slight elevation has commenced to destroy it underground. The surface rivers are too sluggish to cut downwards, but percolating water is in places eating the dolomite away. The Rankin River which flows south-east into the Georgina River, is a big stream during the summer rains, though it sinks to a mere trickle in the dry season. Just across the Queensland border in the Northern Territory it disappears altogether. It has been thought to rise over 10 miles away in a permanent large waterhole known as Bucket Hole in the Georgina drainage. In the Camooweal area certain deep caves, reaching to permanent water more than 200 feet down, have been known the whole century but recent explorations have shown some of them to be very long as well as deep. These caves are formed in dolomite, a carbonate rock in which one calcium and one magnesium atom are combined in each mole-cule. It is not quite as soluble as limestone which is composed of calcium carbonate, with only minor magnesium content; nevertheless fine caves can form in it.

The Mountains of Central Australia

It is a natural though a very long step from the ancient lands of the north to the very centre of the continent, and this would seem a proper sequence to follow in reviewing the whole of Australia's scenery. Such scenery often gains in effect from its unexpectedness. On the long road from Darwin to Alice Springs, though there is much of interest, there is little topographical relief to catch the eye. For seemingly endless miles there is little but dusty plain covered with low bushes and patches of trees. When, therefore, having left the little settlement of Tennant's Creek some 20 miles to the north, the road suddenly enters a region covered with large boulders, interest is at once quickened. We may well pause a moment on our journey to the central mountains to examine them.

These are the Devil's Marbles. From the distance their size is exaggerated, and silhouetted against the setting sun their outlines are varied and weirdly suggestive. The country here is covered with a coarse gravel of quartz, and the boulders may rest directly upon this or upon rounded rocky outcrops of the same material. They may be isolated or piled in heaps one upon the other. Many are almost perfectly round, and they vary in size from a foot or so to monoliths ten or more feet in diameter. To the aboriginal and to the superstitious white they may well appear the playthings of a supernatural giant.

All are of granite, which brings us to consideration of the topography typical of this rock wherever it may be found. Granite may be considered a foundation rock composed of light material, largely silica, lying over the much heavier central mass of the earth itself. When it lies buried deep beneath the extreme outer crust it is plastic, a kind of semi-solid ocean on which the surface rocks float. When through pressure the surface rocks begin to buckle into mountains the granite magma forces its way upwards into the cores of the folds and there gradually cools and solidifies. As it cools its various constituents crystallize, and the slower the cooling the coarser its crystalline structure becomes. It appears on the surface only when the mountains above it are worn away to expose their very cores.

The age of granite may be determined by observing its relations to the surrounding rocks. Where rocks of a known period are intruded by a granite, the granite is of course younger; where overlying rocks lie undisturbed and unaltered the granite is older. By such methods and by radiometric determinations many central Australian granites including those of the Devil's Marbles are known to be Proterozoic in age.

When granite is exposed on the surface it may weather rather rapidly in spite of its great hardness. It often contains within it minute liquid inclusions which assist weathering and thus cause some of the constituent minerals to decompose, particularly the mineral feldspar. This turns to clay, and leaves the small crystals of quartz and mica as sand or gravel. This weathering may be uniform and to a considerable depth. The plain surrounding the city of Bathurst in New South Wales is underlain by decomposed granite, and very little solid rock is left exposed on the surface.

Decomposition of the granite is more commonly uneven, and portions are left quite unaffected. These then remain, becoming gradually rounded with time, and as the softer material disappears they form boulders, often perched one upon the other or balanced precariously in seemingly impossible positions. In higher country boulder-covered hills are typical of granite wherever it is found. Nearly all these localities have their balanced rocks, sometimes so nicely adjusted that they can be rocked by a touch of the hand. Tenterfield and Albury in New South Wales, Mt Buffalo in Victoria and Albany in Western Australia are but some of the innumerable localities where this type of topography is found.

South of the Devil's Marbles lie further great stretches of monotonous plain, but they are a fitting introduction to a wondrous region which lies beyond. Almost in the exact centre of Australia is a great tract of country, about 300 miles from east to west and about the same distance from north to south, which for scenic grandeur must rank as one of the wonders of the world. Until recently few beyond intrepid explorers and an occasional prospector had even glimpsed this region, the approach to which was fraught with extreme hardship and danger. Much of it is still far from the nearest settlement and needs a well organized expedition to reach it, but now any part can be viewed from above in aeroplanes from Alice Springs, and tolerable roads to some of the outstanding features reduce a journey to hours which formerly took weeks by the old-time camel-train.

By accepted geographical rules this should be a desert, but accepted rules must go by the board in this strange world. A desert may be described as a region which from lack of water is uninhabitable. An annual rainfall of below 10 inches is usually accepted as conducive to desert conditions, unless there are springs or rivers flowing from more favoured localities. The annual rainfall here is about five inches, sometimes less, but possibly a little more on some of the higher ranges; moreover it is intermittent, and months and even years may pass without rain falling at all.

The country nevertheless can by no means be described as desert. It has a rich flora, particularly within the mountain gorges. Nor is it uninhabited, for there are many varieties of birds and other animals, and wandering natives find sustenance within its boundaries. There are even cattle stations in those parts where there are permanent waterholes.

The scenery of the region is closely linked with its geology, and both are

intensely interesting. The whole is the eastern part of the wide plateau which
stretches from near the coast of Western Australia more than half-way across
the continent. The plateau is not high, averaging not more than 2,000 feet.
The greater part is an almost perfect plain, but the fantastic mountains of
central Australia rise in places as much as 3,000 feet above the general level.
Of the many ranges and individual peaks three groups may be singled out as
representing the general topography, yet each showing structural differences
to give them individuality.

Of these groups the Macdonnell Ranges run due west from Alice Springs
for over a hundred miles, and with them may be associated such parallel

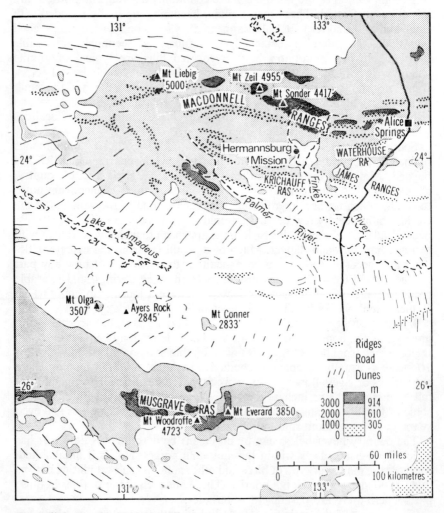

8. Relief map of central Australia.

ridges as Chewings Range to the north and the Waterhouse Range to the south. Quite separate and about 200 miles south-west from Alice Springs are the Musgrave Ranges, a mass of giant monoliths of quite different geological structure. About half-way between the Macdonnells and the Musgraves and rising directly from the plain is a third group of three great, lonely island-mountains. Though separated from each other by long miles of sand ridges and mulga scrub they are structurally so similar that they may be considered together. They are Mt Olga, Ayers Rock and Mt Conner.

To the north-west of Alice Springs are many other ranges and all rising like the others directly from the plain. All are intensely interesting both scenically and geologically.

All the groups of mountains mentioned have features in common, yet each is different from the others. All rise abruptly from the surrounding plains, all are composed of very hard rocks resistant to erosion, and all are approximately the same height. Though all are composed of very ancient rocks, those of the Musgave Ranges are the oldest, and were formed very early in geological time more than 1,400 million years ago, though some slightly younger rocks of highly metamorphosed nature are incorporated also. Mt Conner consists of Upper Proterozoic rocks as well and next in age come Mt Olga and Ayers Rock of somewhat younger Lower Cambrian rocks. Later rocks forming the Macdonnell Ranges were laid down in the Ordovician Period, itself a very ancient one, but young compared with those which went before. The Ordovician Period was the last in which rock-making on a grand scale occurred in this part of Australia. Three stages are thus seen in the building of the foundations of the present topography. Let us take them separately in order of seniority.

THE MUSGRAVE RANGES

To the south of the Macdonnell Ranges and far to the west lies a waterless plain, with long stretches of red sandhills, relieved only by patches of the hardy spinifex and by the silver-grey of the mulga scrub. To the traveller across these dreary plains there appears on the distant horizon a line of rounded hummocks glowing in the early morning sun with a red sheen. These are the Musgrave Ranges. In this land of no perspective they are not imposing from the distance, there is no clue as to their magnitude, and at first sight they may well appear as minor hills but a short distance ahead. To the slow approach of a camel train they may be in view for many days before their base is reached.

Even from a few miles away they do not appear of great magnitude, unless perchance there be some familiar object such as a motor truck in the fore-ground to give perspective. Yet they rise in places to 3,000 feet above the general plain. Mt Woodruffe, 4,723 feet above sea level, is the highest peak, but so rounded are its contours that it is difficult to separate it from the tumbled confusion of its fellow peaks.

The mountains lie in an east-west belt more than 50 miles long, and the width from north to south varies from about four to more than 10 miles. They do not form a typical mountain range, nor even a plateau. There is no main ridge and there are no main valleys. Most of the ranges form a tangled labyrinth of rounded rocky hills of granite, without discernible pattern, in which the compass is the only guide, though in the western parts there are elongated ridges of quartzite. Here and there the plain pushes far into the heart of the hills, and there are wide and level valleys from which isolated hills rise like red islands in a green sea. Here the vegetation is richer than on the plains outside, and belts of thick and luscious salt bush replace the mulga and the spinifex.

Generally the rock is red granite or rather gneiss. The foliated structure of this rock is not always apparent in small specimens, but from the distance it often gives the appearance of stratification to the whole country. Thus in the Musgraves, though the skyline is generally soft and rounded, the foliation of the gneiss in the larger rock masses tends to produce parallel ridges running generally in a north-south direction. Nevertheless most of the mountains have rounded summits with steep and precipitous flanks. Individual peaks may be separated from their neighbours by deep gorges and ravines or fused together in a confused mass.

There are no rivers, unless such lines of drainage as Officer's Creek be so described. This watercourse, after winding for some distance through a plain between some of the higher mountains, emerges to the south. With a few other watercourses of the same kind it may flow for a short time after the periodic rains, but its water is soon absorbed, and it disappears in the sand of the plain. When dry its course may be distinguished by lines of tea-tree and that hardy eucalypt, the ghost gum.

Yet there is water in this seeming desert even in the driest times. When it does rain the water drains rapidly from the bare rocky slopes, and seeps into the sand at their base, or it may collect in deep pools in the shady ravines. Here it may remain for months or even for years, from rain to rain, forming semipermanent or even permanent waterholes. Although most of these are known to the blacks, recent expeditions have found large and deep pools of which even the natives had no knowledge.

The gneisses of the Musgraves are amongst the very ancient rocks of the world but not as old as the Archaean rocks of the original nucleus of Yilgarnia. These younger rocks were welded onto the latter during a Proterozoic mountain-building period. Then a sea separated Yilgarnia from another land, Stuartiana, in the north, and another north-south sea separated it from land in eastern Australia. Yilgarnia was probably a very high and mountainous land, and where the Musgrave Mountains are was just about upon its northern shore. Material worn from these mountains was carried into the sea to the north, and from it were formed the rocks of which Mt Olga and similar mountains were built.

THE THREE LONELY ISLAND-MOUNTAINS

From the highest points of the Musgraves the horizon of the surrounding plain is more than 60 miles away; though sometimes obscured by dust, it is generally clear cut in the dry transparent air. Far beyond the horizon the summits of heights are clearly visible for 100 miles or more. From such a viewpoint the great spaces of the interior are drawn a little closer together, and it is possible to form some conception of their magnitude.

The unique shapes of the isolated mountains make wonderful landmarks recognizable from any angle. Sixty miles to the north is the flat-topped mesa of Mt Conner, and far to its left the rounded dome of Ayers Rock. Still more to the west and fully 80 miles away the fantastic monoliths of Mt Olga appear from the distance like a batch of gigantic eggs, standing on end and apparently leaning upon each other.

Mt Olga is surely one of the strangest mountains in the world. It is not one peak, but a number of enormous monoliths arranged roughly in a circle, with four of the highest side by side on the south-western face. They rise straight from the plain, each with a rounded summit and with smooth joint-less and almost sheer sides. The largest is 3,507 feet above sea level and about 1,800 feet above the surrounding plain. Although quite close together they are separated by narrow vertical chasms, into which the sun can shine but a brief hour or so each day.

Within the chasms is another world. Here drainage is held long after the outside country is completely parched. In the deepest recess it is doubtful if moisture ever completely disappears, even in the most severe drought. The variety of plants is astonishing, and it is from here that most of the 600 species recorded from the area were taken. The bottoms of the gorges are filled with a dense semi-tropical scrub, matted together with lianas and vines, and containing a wealth of animal life, rock wallabies, lizards, spiders, insects and land shells. In the narrowest of the gorges the sides are quite vertical, and go up and up until the sky is reduced to a mere ribbon of blue overhead or completely lost where the walls overhang.

From outside and from any distance it is hard to realize the great height of the precipitous rocks. There is the usual lack of perspective, no large trees nor buildings are present for comparison, and the mulga bushes merge towards the base into a uniform field of silvery green. Even single bushes appear merely as minute dark shapeless specks.

Like the Musgraves the rocks of Mt Olga are red, a brilliant red which gleams in the sunshine. However the rock consists of waterworn boulders of various sizes which have become compacted by a hard siliceous cement into the rock called conglomerate. The hardness of the cement and the enclosed boulders is about the same, as is their resistance to erosion; so instead of the boulders protruding as they do in most weathered conglomerates, they are worn across to a uniform smooth surface, appearing in section like circular tiles on a mosaic floor.

It is probable that the boulders of the conglomerate came originally from the Musgraves or from similar rock foundations to the south. They must have been deposited quite close to the shore, torn by the surf from a rocky coast and rounded by attrition as they were rolled upon the sea bottom. Finally they were cemented together by the finer sand and hardened into solid conglomerate, long ages elapsing before their exposure on the surface.

Viewed closely, the rocks of Mt Olga appear homogeneous, but from a distance, lines of stratification can be seen. These dip at a steep angle. The rocks are of Cambrian age.

Ayers Rock, the second of the three tors, lies to the east of Mt Olga. With Mt Olga it has many features in common, but there are some striking differences. Instead of a number of separate peaks it is one huge monolith, six miles in circumference and 2,845 feet high, rising abruptly about 1,100 feet from the red mulga-covered plain. It is in the form of a huge red dome, the lower slopes steep and precipitous and vertically furrowed into great buttresses, the upper slopes gentler and rising to a rounded summit.

The rock here is of ancient, hard sandstone of the same age as the rock forming Mt Olga and the beds are tilted almost vertically.

Ayers Rock shows some extraordinary effects of erosion. The great buttresses which sweep downwards from the summit are separated by steep but short gullies, with shady recesses filled with luxuriant vegetation, and in three places permanent pools of clear water. The surface of the rock is everywhere vertically furrowed, the furrows coinciding with stratification, and countless rainstorms of the past have found these lines of weakness to wear out shallow channels. Occasional heavy rains give rise to short-lived but magnificent waterfalls down these furrows.

Weathering has also caused much exfoliation of the surface which has in places come away in great slabs. In the north-west corner one enormous slab of rock 20 feet thick lies on a 50 degree slope and is separated for most of its height of 200 feet by a crevice four feet in width. Much of the surface of the rock is deeply pitted with caves due to weathering. Chemical weathering forms a stronger outer crust and a weaker inner layer of rock which crumbles more readily. Winds help to form the caves by removing particles of rock detached by physical and chemical weathering. Caves on the ground level are "decorated with strange totemic paintings from the Old Race, to whom Oolra is a place of awe, where the wind moans always between sunset and dawn".

Far to the east of Ayers Rock but visible from its summit is another isolated mountain, Mt Conner. It has an entirely different appearance from the other two great island-mountains. It is a flat-topped mesa or tableland, oval in plan, about three miles in length from east to west and about one mile from north to south. It also rises over 1,000 feet above the plain. The summit is perfectly flat above an encircling escarpment of vertical cliffs nearly 400 feet in height. From the base of the cliffs a steep slope leads downwards to the plains. On the south side, however, the escarpment is broken by several ravines, which yield rough access to the top.

A nearer view shows the slopes to be composed mainly of talus, masses of rock, large and small, fallen from the cliffs above. The slopes are mostly

Monoliths of Mount Olga, up to 1,800 feet high, Northern Territory. *CSIRO Land Research.*

Advancing sand dune, Northern Territory. *CSIRO Land Research.*

Mountain range in folded quartzite north of the Simpson Desert, Northern Territory. Elongated sand dunes end against its foot. *J. Cavanagh, CSIRO Land Research.*

Rumbalara Hills, Northern Territory. Flat-topped "tent" hills are due to erosion of former plain with a hard siliceous capping of fossil soil. *J. Cavanagh, CSIRO Land Research.*

bare, but here and there trees of white cypress pine and ironwood have found precarious root. A dense growth of spinifex covers the flat top of the mesa.

The main difference between Mt Olga and Mt Conner is geological. Whereas the former is composed entirely of conglomerate, only the upper portion of Mt Conner is composed of this rock. Below the escarpment are beds of fine siliceous sandstone, which, like the conglomerates, are perfectly horizontal bedded. The sandstone, though hard, does not resist erosion quite so well, and thus it undermines the conglomerate, which is continually breaking away and adding to the talus slope below. It is a story of slow but constant destruction. The cliffs are kept vertical, or they may even overhang as much as 50 feet in places. Great masses of rock are balanced precariously and seem ready to fall at any moment. Deep crevices penetrate far into the hillside, and the lower part of the cliff is honeycombed with caves and rock shelters.

THE MACDONNELL RANGES

To complete the story of the central Australian mountains the Macdonnell Ranges come as a fitting climax. Geologically they differ considerably from the Musgraves and from the isolated mountains to the south; there is a greater variety of rock formations, and the strata are so arranged as to give the mountains a definite topographical pattern. They conform more to the conventional idea of mountains, for there is a main axis or rather axes, the main ridge running in an east-west line for about 200 miles, with secondary parallel ridges both to the north and south.

Though lying in almost the exact geographical centre of Australia the Macdonnells are the best-known and most accessible of all the central ranges. Explorers came to them early and found a temporary refuge from the waterless country north and south. There was water here and there in the rocky gorges intersecting them from north to south, there was shade, and there was even game in the rocky foothills. The railway now passes through the eastern end, and the thriving township of Alice Springs nestles at the foot of the northern slopes. There is even a road winding through the mountains for some distance to the west and then skirting their southern border to the lonely Hermannsburg Mission some 70 miles from Alice Springs.

Considering their extent, the ranges are not high. They average only from 1,000 to 1,500 feet above the general level, which is itself 2,000 feet above sea level. There are some higher peaks. Mt Conroy, 30 miles west of Alice Springs, is 3,726 feet high; much farther west is Mt Sonder, 4,417 feet. The highest peak in central Australia is Mt Liebig, 5,000 feet, at the western end of the Macdonnells. Indeed much of the western Macdonnells is yet unsurveyed. Sufficient of the eastern part is known nevertheless to reveal most of the story of their birth, their growth to the majesty of a high alpine range, and their old age, when a mere fragment of their former grandeur remains.

There is this to be said of mountains in arid country, the absence of dense vegetation on their heights and of soil in their valleys reveals in stark nakedness every detail of their structure. The rocks of the Macdonnells belong to several periods, each contributing something to the present topography. Below all is a pavement of the most ancient rocks, principally gneisses similar to those forming the Musgraves, and above these are conglomerates and quartzites deposited in the Proterozoic sea when Australia was divided into several islands. Above these again are Cambrian rocks laid down in the sea in which the rocks of the Barkly Tableland were also deposited. Deposition continued without interruption into the Ordovician Period. This sea did not extend right across Australia, but covered western Tasmania, most of Victoria, southern New South Wales and the centre of Australia. A long gulf near its head ran east and west, and in this the uppermost rocks of the Macdonnells were laid down.

These rocks consist of sand which has since hardened into thick beds of compact resistant quartzite and of mud which is now shale and slate, comparatively soft and subject to erosion. It is the relative resistance of the rocks to erosion which has shaped the present topography.

These rocks were involved in violent earth movements after the close of the Ordovician Period, when the rocks were exposed to intense pressure from the north and south, and squeezed upwards into a series of parallel folds and troughs, running mainly east and west. There was also some pressure from the east and west, and in places the strata were elevated into great domes or depressed into circular basins instead of being pushed into folds and troughs.

It was at this stage that the Macdonnells probably attained their greatest splendour. They were then a mighty range of mountains towering from 10,000 to 15,000 feet into the sky, but like all the land of their time they were more desolate than any at the present day. There were living things in the sea, but from all the evidence life had not yet come to the land. Though rivers ran down the valleys, there were neither trees nor bushes, no grass, no ferns, not even a lichen or moss upon the bare rocks, and without vegetation there was of course no animal life.

There are some folded Devonian conglomerates in the Macdonnells, which tell of some later less intense earth movements. But after this the story of the Macdonnells is the story of all central and western Australia. For age after age and for period after period there were no earth movements of any magnitude, nothing but the slow relentless erosion of the higher land. The mountains became lower and lower, and the material from their destruction was carried by rivers to distant oceans, or filled the valleys and levelled the plains. There came a time when all the mountains had disappeared, when most of Australia was practically level, the only hills a few low mounds or residuals of the very hardest of the rocks.

This phase lasted for many ages to a time still remote from us but belonging to the later geological periods. It is difficult to fix the exact time when the western portion of Australia again began to rise, but it probably did so during or after the depression in the Cretaceous Period when the sea overflowed from the Gulf of Carpentaria to South Australia. This rise was not quite uniform throughout, but its general effect was to convert the western half of Australia into a vast tableland with a maximum height of slightly over 5,000 feet. The rise was gradual and probably continued right through the Tertiary Period.

Today there are many gorges cutting completely through the Macdonnells, gorges once occupied by streams to join the mighty Finke River on its way to the south. The gorges form natural passes through the mountains from north to south, and within 60 miles of Alice Springs there are at least seven of them. It is conceivable that the original Finke River did not at first flow directly south to the Southern Ocean, but that in Cretaceous times it turned eastwards into the Cretaceous sea.

The development of the river system here is complex. Much of the system goes back to the retreat of the Cretaceous sea when upwarping of a gentle erosion surface set rivers flowing southward across the rock strike of the country. Belts of resistant rock were etched out by erosion and broad, gentle gaps cut through them. Later rejuvenation caused the rivers to cut steep gorges within these gaps; the strike ranges heightened and their flanks steepened as longitudinal vales between were developed. In this way the transverse drainage discordant to structure was inherited from the older surface represented in the flat summits of the ranges. Not all the gorges formed in the same manner however. Some of the finest such as Standley Chasm and Spencer Gorge in the Chewings Range do not transect a ridge of hard rock completely. They are still in process of doing this. These courses are not inherited but due to headward erosion by southern flank rivers which drain to a lower plain than do those on the north side. Some other gorges through the Chewings Range such as Ormiston Gorge are due to river capture.

All the mountains, the Musgraves the great island-mountains, and lesser ranges rising from the plain some hundreds of miles to the north are residuals of the harder rocks, their shapes dependent upon their individual geological structure. Water has been the main agent lowering the land between them but in the Pleistocene and in the Holocene there have been oscillations between wetter and drier conditions. In the drier phases wind has been important, not as an erosive agent but because of its transport and depositional work. Recent investigations indicate that there were two main phases of dune formation, both of Pleistocene age.

In the Macdonnells the alternation of hard beds of quartzite with softer shale has produced some remarkable topography. In the main range the axis of the folding is due east and west, and the hard quartzite stands up as long undulating ridges, one above the other, each ridge with a steep escarpment on its inner edge, showing where a bed of shale has formed a line of weakness. There are many east-west valleys within and parallel to the main range, each showing the position of softer rocks.

To the south of the main range the topography is still more striking. The Krichauff Range has been worn to a lower level than the Macdonnells, and its quartzite ridges stretch for mile upon mile in tilted parallel lines, to which former cross erosion by water has given a curious undulation, referred to locally as "tombstones".

The Waterhouse Range is the most curious of all. Here the folding of the rocks was in the form of a lofty dome, its longer axis from east to west. The mountain conforms almost exactly to the original geological structure. It still remains an oval dome, though its top has been completely worn away and is now only a little more than 1,000 feet above the plain. The hard layers of quartzite completely encircle it, and all dip outwards, so that they conform nearly to the angle of its sloping sides. The truncated top, composed of softer rock in the core of the folding, has been slightly hollowed out to a plain many miles across and actually slightly lower than the surrounding rim. Here is a little isolated world, almost like the broad crater of an extinct volcano. There is not much life, but bushes mark where a usually dry watercourse meanders through the full length of the centre, and where numerous tributary watercourses join it from either side.

Each hill in this region has its own individuality. There are fossils to be found in the rocks, extinct shells and crustaceans that had once lived abundantly in the Ordovician sea. For the naturalist there are many strange animals and plants, all specialized types adapted for life in an arid climate. There are minerals in the hills, and hardy prospectors have added much to knowledge of the remoter parts. There is even rain at times, and the watercourses run and rivers temporarily resume their former importance. Then the whole country for a while blooms again. And there are clear bracing air, wonderful sunsets, and pervading all, gorgeous colouring, deep mauves and purples, yellows and heliotropes, and a vivid striking red.

Southern Australian Landscapes

THE NULLARBOR PLAIN

Travelling from Western Australia for about 160 miles east of Kalgoorlie, the train enters one of the most remarkable physiographical regions in Australia, the Nullarbor Plain. For long miles before this the country is a plain, sloping it is true slightly downwards to the east; but there is little to show, beyond the absence of belts of the salmon gum, where the ancient land surface of Yilgarnia ends and a much newer land commences. The country is now treeless, covered here and there with low patches of salt-bush, without watercourses, the only soil a thin covering of wind-blown sand rising occasionally into low dunes. It is only where occasional outcrops of rock break the surface that the change is revealed. Where before had been the granites and gneisses of the Pre-Cambrian periods is now horizontally bedded limestone of the Miocene Epoch, laid down during the very last interlude when any considerable area of Australia was submerged beneath the sea.

The railway traverses the old bed of this sea for about 400 miles, 330 miles in an absolutely straight line, the longest straight stretch of railway in the world, and on an average about 100 miles from the present coastline. The plain continues northwards as much as 100 miles before it is replaced by the great dune-covered deserts of the interior. Throughout its whole area there are only rare traces of stream or watercourse where surface water may have at one time run.

When Eyre crossed the Nullarbor Plain in 1832 he travelled from east to west and followed a more southerly route than the railway. His way lay along the coast, sometimes on the beaches at the foot of the cliffs, or, where the cliffs rose directly from the sea, along the level ground at their top. He found the country here much the same as farther inland, except that the scrub was denser and difficult to penetrate, and there were many small stunted trees. The coast was very striking, a continuous escarpment from 150 to 300 feet high, bordering the whole of the Great Australian Bight. This escarpment impinges right on the sea for most of its length, but in the centre it recedes slightly, and from Eyre to Eucla there is a narrow coastal plain.

The story of Eyre's terrible journey is written in Australian history. He found grass occasionally for his horses, but it was water that he consistently sought, for without water he could not continue, when indeed he must continue or perish, since he had proceeded beyond the point whence he could still return. Eyre found water, since first, he was a good bushman and had learned much from the natives, and secondly, the Nullarbor Plain, though arid, is really not a desert. Before the coming of the white man it had supported a small nomadic native population. There is a small regular rainfall which does not run off the surface but sinks rapidly into the permeable limestone. In our later chapter on caves mention will be made of the strange underground lakes being discovered there one by one, but it was not from these that Eyre found his water. Here and there along the coast are patches of sand dunes. Many of these had been formerly observed from the sea and their positions recorded, and to Eyre they were the widely separated oases which made his journey possible. In the hollows between the dunes water was generally discovered by digging to a depth of about five feet, and though it was sometimes brackish, by a fortunate chance it was fresh, cool and abundant when his need was greatest.

The plain, though dreary scenically, is of extreme geological interest. It marks where a large section of Yilgarnia, the land which had survived so long, had subsided beneath the sea. This subsidence took place in early Tertiary times or even before, and attained its maximum in the Miocene, when a large part of southern South Australia and Victoria was also submerged. In the western part of the subsidence the sea penetrated deeply, and its shore lay far to the north of the Great Australian Bight. In it lived countless marine organisms, and their remains slowly accumulated into beds of limestone. These have been proved by bores near the present coastline to be over 900 feet in thickness. Fossils are common everywhere in the limestone, and Eyre found time on his journey to remark on the presence of sea shells in the cliffs, and to comment on the former existence of the sea over this part of the land.

When the sea bed finally rose again, probably towards the end of the Tertiary Period, the uppermost layers of limestone formed the surface of the Nullarbor Plain. This was then much as we know it now. In the subsequent interval underground waters have been slowly but inevitably dissolving it away. There are relics of river valleys reaching out into the northern part of the Plain from the Great Victoria Desert where they are crossed and blocked by sand dunes. These valleys probably formed in colder periods of the Pleistocene when greater effective precipitation caused more runoff and streams reached farther over the limestone before sinking underground.

The extraordinary continuity and regularity of the cliffs of the Nullarbor Plain coast prompted earlier geologists to postulate faulting to explain it. However the evidence is against the presence of such faulting. The cliffline is due to the wave attack of the sea operating on horizontally bedded limestones changing laterally very little and to the absence of surface drainage which would make breaks in the cliffs through their valleys.

A FOSSIL LANDSCAPE

We have spoken of the ancient land surfaces of Australia, areas which have existed much as they are now for countless ages. Rightfully these have been looked on with wonder, for in a world of change scenery is the most ephemeral of things, generally fated to destruction from the moment it is shaped. Sometimes scenery is not destroyed but buried, whole landscapes with hills and valleys sinking beneath the sea, or becoming covered with lava-flows until they are levelled into a plain. Sometimes, although very rarely, such landscapes reappear, and stripped of the rocks beneath which they were buried, reassume their place in a new landscape much the same as they were ages before. From even a fragment of such a landscape it needs little imagination to form a much wider picture and mentally to reconstruct the scene as it appeared in a remote age.

In the extreme south of South Australia is such a fossil landscape, now reappearing here and there to tell the story of the past. Nature for once is acting in the role of a picture restorer, carefully removing successive coats of old varnish to reveal the rich tones of an old master. The layers of varnish are marine rocks laid down in the Miocene Epoch, the picture beneath is an old hilly landscape overridden by an ice sheet in the Permian Period over 230 million years ago.

Only here and there fragments of the old landscape are so exposed that they may be said to be part of the existing scenery. Odd patches appear on the extreme southern tip of Yorke Peninsula, but here they are mostly hidden by wind-blown sand or a coating of soil. At Kangaroo Island the old landscape is buried deep and has been reached only by bores at a depth of many hundreds of feet and well below existing sea level. However on the south coast there is striated surface exposed by coastal erosion, which looks as fresh as if the glacier had scratched the rock only a few decades ago.

It is in the southern part of Jervis Peninsula, from St Vincent Gulf across to Victor Harbor, over an area of about 100 square miles, that the original Permian landscape is frequently visible. Here in the Inman Valley only the tops of the ancient hills are exposed on the surface, for the valleys were deep and now go down well below sea level. A bore at Black Valley Creek shows that the old valley floor is now 830 feet below the surface. Nevertheless, as these valleys were filled at the time by glacial débris in the form of boulders and clay, the greater part of the present surface of low relief may still be said to be part of the Permian landscape as it was at the conclusion of the period. When the tops of the hills are first exposed to erosion, they retain the detail of former glacial erosion, broad pavements of very ancient rocks, still highly polished and deeply scored by the ice sheet which passed over them so many millions of years ago. However they soon begin to be modified by weathering and erosive forces so that, though their broad form may be inherited from that ancient time, their detailed nature ceases to be glacial.

Rosetta Head, a prominent rocky knoll on the coast west of Victor Harbor, is a hill which has preserved its shape since Permian times. It is a rounded prominence composed partly of granite, partly of ancient Cambrian sediments, and its surface is still polished and scored by the ice sheet which passed over it. About three miles north is another Permian hill, Crozier Hill, also preserved from the past. Between them is a flat, scrubby area, composed of old glacial débris, and above this, in at least two places, the tops of other Permian hills are visible. Scattered everywhere through the scrub are erratics or ice-borne boulders showing typical glacial scratches and polishing. Some of the boulders are very large, and one of granite near Encounter Bay is 23 feet in length.

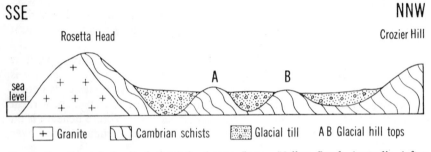

9.   Section through fossil glaciated landscape, Inman Valley, South Australia (after Howchin).

Glimpses of a similar fossil landscape are revealed in Victoria many hundreds of miles to the east. The evidence of its wider extent is clear, but it is generally too deep beneath later rocks to be an actual part of the existing landscape. The best locality for its study in Victoria is in the Bacchus Marsh district, particularly along the gorge of the Werribee River a few miles above the town. Here in one place is revealed in section as it crosses the gorge, a complete typical U-shaped glacial valley, filled with glacial boulders and clay, its bottom where exposed polished and scratched by the ice.

On Rosetta Head we stand on a fragment of Permian landscape almost as it existed 230 million years ago. But what a different vista there must have been then. Now far to the south the ocean stretches from our feet to the far Antarctic continent. To the east and to the west the Southern Ocean completely encircles the globe, except where South America pushes a long tongue southwards into higher latitudes. In Permian times it was not thus. Not only was the South Pole very close to the south in latitude 45° but there was also land to the south, part of Gondwanaland. Formerly it was thought that this land had sunk beneath the ocean but modern evidence from the ocean floor supports the idea that the land surfaces making up Gondwanaland have drifted apart to give the present pattern of continents. There are similar Permian glacial deposits near the South Pole of today in Australia, in South Africa, in South America and in southern India, all scattered fragments of the former Gondwanaland.

Limestone cliffs of the Nullarbor Plain at Koonalda, South Australia. *J. N. Jennings, Australian National University.*

Glacially scratched granite, Kangaroo Island, South Australia. Effects of Permian glaciers 230 million years ago now exposed by erosion. *J. N. Jennings, Australian National University.*

Cooper Creek filling Lake Eyre in 1949. *Commonwealth Department of Supply.*

Hard, dry surface of a salt lake, South Australia. *South Australian Department of Mines.*

In the chapter on the Kimberleys the Permian climate was spoken of as cold, marking as it did one of the great glacial ages that have at long intervals apart visited the world. In all there are four extreme glacial ages known, one in the dim Pre-Cambrian times, one in the Cambrian Period, the third in the Permian Period, the fourth, the one we call the Great Glacial Age, only recently ended and through which the human race evolved from his somewhat ape-like ancestors.

Glacial ice leaves on the face of the land very definite evidence of its passing, evidence which may remain ages after the ice has melted. In periods when the world's climate is cold, snow falls and consolidates until whole continents are buried beneath a sheet of ice. The ice sheet may be thousands of feet thick like that covering Greenland and Antarctica at the present day. Squeezed outwards by its tremendous weight the ice flows over the land, abrading the heights, gouging out valleys, and polishing the surfaces of even the hardest rocks. Much of the wear is due to fragments of rock embedded in the bottom of the ice which act as chisels and leave characteristic scratches as they pass. Sometimes these scratches deepen as they lengthen down the movement of the ice so that the single direction of the ice sheet's movement can be inferred. However it is not always possible to infer the ice movement as a single direction; uniformly deep scratches offer two possible directions 180° apart.

There are many such ice-striated pavements in South Australia, Victoria and Tasmania belonging to the Permian Period. In South Australia the scratches show that the ice flowed from the south, in Victoria it came from the south-west, and in Tasmania it came from the west. This was land ice, for no sea ice ever overrode the land to this extent. Moreover, it must have come from high land, for ice like a river moves downhill. Thus the former presence of continental land south of Australia is clearly demonstrated, and even something of its construction may be learned from the material torn from the land as the ice passed, to be left far from its source when the ice melted.

SOUTH AUSTRALIAN MOUNTAINS

Adelaide is a fair city set on a plain by the sea. Across the few miles of plain to the east and visible from the city are the slopes of the Mt Lofty Range. These hills are the green heart of the State, and they stand like guardian sentinels over the destiny of the white man's enterprise. A godmother bringing gifts would perhaps be an apter simile, for without this higher land there would here be but another part of the long arid coastline stretching far to the west.

On the rainfall map of South Australia the 10-inch isohyet west of Adelaide approaches closer and closer to the sea, until in the Great Australian Bight it is right upon or even beyond the shore line. Within it is another line, the 5-inch isohyet, and beyond are regions with a rainfall so low that they are classified as desert. About the Mt Lofty Range the isohyets run in a series of concentric circles, the figures rising higher and higher until that in the centre is 40 inches. The life-giving fluid from these hills made possible the creation of a state, and reservoirs in these hills used to be Adelaide's main water-supply. Now this city and other South Australian towns are chiefly dependent on pipelines bringing Murray River water across the Mt Lofty Range.

The term mountain or range is a relative one. What are called foothills in the Himalayas would be lofty mountains in many other lands and would certainly overtower the Mt Lofty Range by many thousands of feet. The importance of the South Australian mountains is not reflected by their mere height above sea level. The general average is only about 1,500 feet, with a few peaks above 2,000 feet, including the highest point, Mt Lofty, 2,384 feet.

It is a pleasant and picturesque region, this chain of hills so close to the city. It is not very wide, and the train to Melbourne puffs through its valleys for about 20 miles with no excessive grades before emerging to the plains of the Murray River Valley. The rivers are short but perennial, fed by numerous springs, the reservoirs of the rain falling mostly in the winter months. The Torrens River flows almost directly west to St Vincent Gulf, the Onkaparinga River flows south through the range for many miles before it too breaks through to the west. These and other streams have excavated for themselves steep-sided gullies, rarely more than 500 feet deep, and seldom narrowed sufficiently to be called gorges. There are no spectacular waterfalls and few prominent cliffs or escarpments. Forests of South Australia blue gum (known in other states as white ironbark) grow on the rounded hills, stringybark and peppermint upon the slopes. In the gullies there is often a dense undergrowth of banksias, wattles, hakeas and other typical plants, and many wild flowers grow on the hillsides.

In the bottoms of the valleys the soil is rich and well drained, and prosperous farms and orchards are found even in the most secluded spots. Here, even in the scorching days of summer, is pleasant shade to be found, and in the winter there is shelter from the cold blasts sweeping down from the arid interior.

The ranges are composed of very ancient rocks, but as they stand they are not very old. The actual age of a mountain is from the time when it was last elevated appreciably above sea level; its actual components may have originated long previous to this. The story really begins before the Cambrian Period, when Australia was divided into two islands, a period already discussed when we spoke of the Barkly Tableland. In this ancient sea many thousands of feet of sediment accumulated over most of South Australia and this process continued into the Cambrian Period. Beds of sand later hardened into sandstone and quartzite, mud into shale and slate, and beds of limestone were formed by the growth of reefs of extinct sponge-like organisms called *Archaeocyathinae*. Early in the whole development there was a great glacial age, or rather two glacial ages, separated by a long interval of time. These were the earliest known of the several great ice ages in the world's history. During this glaciation large deposits of débris were distributed by floating ice over the bed of the ocean and accumulated on the neighbouring land.

At the end of the Cambrian Period a great change took place. The newly-formed rocks were subjected to enormous lateral pressure. By this they were squeezed from east to west and pushed upwards into a series of great folds. Faults, thrusting blocks of rock up over others, also developed. A high range of mountains was thus formed, stretching for hundreds of miles to the north and probably to the south also. This range persisted for a long time, but was gradually worn away until it was reduced to a nearly flat rocky platform very little above the level of the sea. So it remained, age after age, period after period, at such a low level that it was spared further destruction until comparatively recent times. In early Tertiary times there came the general subsidence in southern Australia which submerged much land beneath the sea, the same subsidence as was responsible for the rocks of the Nullabor Plain. The higher parts of the Mt Lofty Range escaped submergence but was surrounded by the new sea in which were deposited many hundreds of feet of sand and mud containing marine shells, the bones of whales, the teeth of sharks and the remains of many other marine creatures.

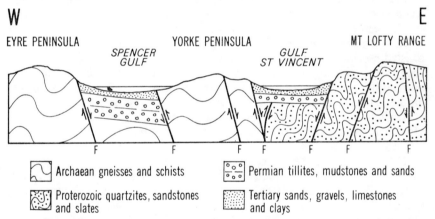

10. Section across the rift valleys of Spencer Gulf and St Vincent Gulf and the horst of Mount Lofty Range.

Though it began long before in the Tertiary, the last stage continued through the more than 2 million years long Pleistocene Epoch and persists to the present day. The earth movements which produced the present topography were somewhat complicated. There was over all a general elevation of the land so that the sea receded to approximately its present position; but the elevation was not everywhere equal, and it was followed in some places by another subsidence.

During the general elevation the Mount Lofty Range came again into existence, the elevation being greater and continuing longer than in the adjoining regions both to the east and the west. Technically the range is a horst, a large block of country lifted above the surrounding levels, its margins defined by great fractures or faults.

It was probably in the later stages of the elevation that the western portion again began to sink, the boundaries between the rising horst and the sinking area being defined by great north-south faults with a total vertical displacement of about 4,000 feet. Thus were produced the great rift valleys which are traceable to the north as far as Lake Torrens, and which allowed the sea once again to penetrate some distance into the land. Both Spencer Gulf and St Vincent Gulf lie at the southern end of these rift valleys.

Though elevated above the sea to a height even greater than at present, the Mt Lofty Range was still to suffer modifications to bring it to its present form. Its original appearance was probably that of a disjointed tableland, descending to the east and west in a series of tilted blocks, each marking one of the faults along which the land had risen. Parts were capped with a thick deposit of Tertiary marine sediments. All through the Pleistocene and to the present day erosional forces have been at work. In Pleistocene phases of greater run-off, erosion would have been even more rapid than now. The first rocks to disappear were the soft superimposed marine beds, and of these little trace now remains. In the Hindmarsh Tiers there is an outcrop of these rocks 900 feet above sea level, but most of the remaining outcrops are fragmentary and small, often only a few yards in extent.

The very old Proterozoic and Cambrian rocks beneath the Tertiary have resisted erosion better, but even these are being relentlessly worn away and the general level lowered. Many of the valleys cut by the streams run north and south for some distance, following the strike of the softer strata, while the hardest rocks remain as the main hills. The process is probably slowing somewhat, partially due to phases of drier climate in the Pleistocene and Holocene, and partially to the gradual lowering of the land and the lessening gradient of the streams.

The topography north of Adelaide is for some distance very similar to that in the south. The horst of the Mt Lofty Range continues, and the roads and railways lie within its western margin, following fertile valleys flanked by wooded hills, frequently higher on the eastern side owing to the original step-faulting of the land. The tableland here is much dissected, the valley bottoms are broad, the hills rounded. Rivers similar to the Torrens are crossed, the Gawler, the Light and the Wakefield, all running to the west, but fed by tributaries flowing north and south and following the strike of the Cambrian strata.

As the head of St Vincent Gulf is passed the country slowly changes, though its structure remains the same. The vital factor in the scenery is the rainfall. Beyond St Vincent Gulf there is a belt of country about 40 miles long with an annual rainfall of 20 inches, and behind Port Pirie on Spencer Gulf there is a small area with 25 inches. North from here it becomes lower and lower.

Even with this rainfall the country differs greatly from that in the south.

About Burra, only 100 miles north of Adelaide, the tableland averages about 1,600 feet, rising in a series of parallel north and south ridges to 3,000 feet at Mount Bryan. This is a wind-swept region, with low winter temperatures, and is practically destitute of trees, the gum forests of the south having nearly disappeared. Only the casuarinas seem capable of withstanding the conditions and they cling to the most exposed positions on the hillsides and quartzite ridges.

Just where the Mt Lofty Range passes into the Flinders Ranges is not well defined, but the limit of 20 inches rainfall may be taken as the boundary. Thence for some 200 miles the horst continues to its northern extremity, where the rainfall is 10 inches a year. Even this is higher than that on either side, where parts are true desert, the salt-encrusted surface of Lake Frome on the east and Lake Torrens on the west. The northern tip of the Flinders Ranges finally dip down beneath vast plains, now a veritable desert of gibber plains and seas of sand. In certain colder stages of the Pleistocene these plains were more fertile than at present and there were rivers when precipitation was more effective, if not necessarily greater.

For the last two or three hundred miles the Flinders Ranges differ more and more in appearance from the Mt Lofty Range. The peaks are higher and bleaker, rising generally direct from the flat surface of the great rift valley which ever parallels them to the west. Their summits consist of bare ridges of rock, mainly quartzite standing sharp and clear-cut in the dry air. The valleys are narrow rocky gorges, the streams intermittent. Vegetation is still thick in the gullies, but many of the heights are quite bare, even the hardy cypress pines and casuarinas failing to find sufficient moisture for their existence.

Some curious topographical features are caused by the structure of the original Proterozoic and Cambrian rocks. In the south these were folded by pressure that come from the east and the west, and they were squeezed into a series of arches and troughs running from north to south. In the Flinders Ranges the folding was more complicated, and pressure came from the north and south also, resulting in the formation of large domes and basins. After much erosion only the plan of the domes is retained by the hills; the harder layers stand out in series of concentric steps, the outward slopes gradual but the inward slopes steep. The basins remain as huge natural amphitheatres encircled by tier above tier of rocky ledges. These basins are known locally as pounds.

Some of the domes such as that at Blinman are not due solely to upward arching of the beds. In addition weaker rocks underneath have been squeezed up into the core of the dome rather like a thick dough. This commonly happens when the rocks below consist of rock salt or gypsum which are more easily extruded in this way than most sedimentary rocks.

One of the most extraordinary of the basins is Wilpena Pound, lying in the heart of the ranges about 80 miles north-east from Port Augusta. Rising from a complicated mass of lower hills is an isolated plateau, oval in shape, some 12 miles long and nearly five miles wide. Surrounded entirely by a steep escarpment of hard quartzite, it is not flat on top like most plateaux, but slopes inward to a hollow basin, like the crater of a vast volcano. The inner basin conforms exactly to the dip of the ancient strata composing the basin. The outer rim is a steep ridge with a serrated skyline, and it rises in the north-east to St Mary Peak, 3,900 feet high. The only easy way into the pound is through a narrow gorge on the east side, down which Wilpena Creek, when running, flows towards Lake Frome on the plain. Unlike many parts of the Flinders Ranges the centre of Wilpena Pound is well covered with vegetation, notably the sugar gum (*E. cladocalyx*), acacias, and in good season fields of grass.

In the northern part of the Flinders Ranges there are other pounds, though none is as well defined or as picturesque as Wilpena. There are other areas where the Proterozoic rocks stand on edge to form serrated ridges separated by deep gullies. In still other parts the strata are horizontal, and though much dissected by gullies, the original tableland is more clearly revealed.

In the extreme north, that portion known as the Gammon Range is of this type. The backbone of the range is here over 3,000 feet above sea level, and consists of a flat-topped ridge running to the north-west. The ridge, though flat, is deceptive, for it is intersected by innumerable rugged gorges, many of them flanked by cliffs hundreds of feet deep. The Gammon Range makes a great contrast with Wilpena Pound because its broad top is a gentle dome of the same quartzite which in a structural basin forms the Pound.

A curious feature of the Gammon Range is a mysterious booming sound that occasionally comes from the heart of the mountain to the outside world. It was known to the natives long before the coming of the white man, and in their legends is attributed to the great snake Arkaroo. Arkaroo once drank Lake Frome dry, and then retired to the mountains in a chronic state of internal discomfort. It is the groans of Arkaroo that periodically disturb the solitude. There has been much speculation about the origin of the sounds, but they are usually attributed to the falling of rock masses from the cliffs into the gorges below.

Though in an arid belt of country the northern Flinders Ranges are certainly not desert. On weather maps the 10-inch isohyet is shown as a line encircling it, and though records are not available, it is probable that in parts the rainfall is over 15 inches. Rain mostly falls in winter, and though the streams cease to run in summer, there is usually water in the beds of the gorges. There is on the whole a surprising amount of plant and animal life. Spinifex on the outer edges gives place to cypress pine, and the slopes and some of the mountain tops are covered with a thick scrub of tea-tree, casuarina and patches of mallee, with here and there the trunk of a blackboy. In the gorges are large eucalypts and many other plants.

THE GREAT SALT LAKES

Around the northern tip of the Flinders Ranges and surrounding it like a gigantic horseshoe is one of the most desolate regions in Australia, the region of the Great Salt Lakes. In 1949-50 a great flood of freshwater poured into Lake Eyre for the first time in the memory of European settlers. The white man has been in Australia for less than 200 years, and our knowledge of the vast interior is much shorter than this: hence, except by inference, we know little of the incidence of such floods.

When the first explorers attempted to cross the continent from south to north, the salt lakes were thought to form a continuous, impenetrable barrier athwart their path. Gregory passed south between Lake Eyre and Lake Blanche and Warburton south between Lake Torrens and Lake Eyre prior to a successful crossing; nevertheless Stuart by-passed the lakes to the west on his great journey.

In 1949-50 after long rains, rivers that seldom flow brought floods of precious water across the continent from the highlands and plains of northern Queensland. What was previously a vast depression of salt-encrusted mud was once more a lake, covering hundreds of square miles. But by 1953 the lake was dry again and though some flood waters have reached the lake basin since then on one or two occasions, it has not filled up once more. It is estimated that under present climatic conditions, Lake Eyre may fill up once or twice a century.

As Lake Eyre usually appears, it is a dreary and monotonous region, classified as absolute desert. It stretches from horizon to horizon, a vast plain depressed 52 feet below sea level. From the south the way to it runs through red sand hills covered with salt bush and acacia, merging gradually into a plantless expanse of salt-covered mud. The white surface glistens in the sun, relieved here and there by black patches of the hardened mud lying everywhere beneath. The air shimmers with perpetual mirage. From a short distance it appears a sea of water, the shores where visible lifted high above the horizon. Across the narrow part of the lake a distant hill may be 40 miles away, but is clearly visible floating in the sky with clouds both below and above.

Lake Eyre is over 50 miles from north to south and nearly as wide, and it covers over 3,000 square miles. South Lake Eyre is similar but smaller than the main lake, it is connected to it by a narrow channel, and has an area of 400 square miles. To the east the country is chequered with sandhills, salt-pans and spinifex, and is broken only by the wide barely discernible channel of Cooper Creek. Ordinarily Cooper Creek is a river in name only, for its bed is usually dry, and it is only at long intervals that heavy rain in far north Queensland makes it for a brief time a raging torrent. Even such floods as these rarely reach far into the bed of Lake Eyre, for they spread out and are rapidly absorbed by the sandy soil upon its margins.

Farther to the east and around the tip of the Flinders Ranges are other dry lakes, similar to Lake Eyre but smaller. They run in a nearly north-south arc, Lake Gregory, Lake Blanche, Lake Callabonna and Lake Frome. North of Lake Eyre the country for 100 miles is covered by innumerable salt pans, and north again is the Simpson Desert, a sea of sand, where the dunes run in parallel north-south waves from horizon to horizon.

The land to the east is not quite waterless, since though it has but an annual rainfall of about 5 inches, it lies on the fringe of the Great Artesian Basin. Beneath the plains is the story of another age, for this was once the south-western shore of the Cretaceous Sea. When the land sank in the Cretaceous Period, the sea flowed into Australia through the Gulf of Carpentaria, and spread far across the continent for a time making a strait through to the Southern Ocean. Rocks laid down in this sea have been penetrated by numerous bores, and have yielded sea shells and the bones of long extinct reptilian monsters. Right to the foot of the Flinders Ranges there are scattered and lonely sheep stations, dependent entirely upon the precious artesian water tapped by the same bores deep below the surface.

From the scenic point of view salt lakes have few attractions, except perhaps clear air and unlimited elbow room. There is more of interest to the geologist, the geographer and the naturalist. The story through later ages is one of changing climate and of slight earth movements which entirely altered the drainage. There were times when it was better watered than now, with more vegetation and animal life.

Recent geological work in connection with the search for oil and gas has given us a better idea of the history of Lake Eyre, which was in existence much earlier than was previously thought. In early and middle Tertiary times, ancient plains in these parts were subjected to pronounced weathering. A hard siliceous surface developed as a result of the seasonal movement of water in the soil. However bores tell us that this siliceous crust is absent beneath Lake Eyre; this implies that a lake already existed there. Then in Miocene times the area was affected by gentle earth movements, warping up a series of low domes around the Lake Eyre basin but depressing the lake bed itself. On the western side a north-north-west fault dropped that flank more so the depression became asymmetrical. In this enlarged depression, in a brackish water lake, limestone and dolomite formed and sands and silts were washed in from the higher ground around. In deposits from this stage in the lake's history the bones of ancestors of today's marsupials have been found at several places west of the present Lake Eyre. This very large lake was present in later Tertiary times, in the Miocene or Pliocene epochs.

In late Pliocene times, the whole area was uplifted and erosion attacked the domes around the lake basin especially. The products of this erosion were carried out of the area by rivers flowing southwards to the sea, probably through the Lake Torrens depression. This lasted well into the Pleistocene. Then once again it ceased to drain so effectively to the sea; gypsum crusts formed over the alluvial plains of the basin. Also artesian springs fed pools in which freshwater limestones accumulated. The bones of the extinct marsupial, *Diprotodon,* have been found in deposits of this time. *Diprotodon* was a giant wombat-like creature about the size of a rhinoceros. Radioactive carbon in these bones and in brackish water snail shells have given us an idea of the age of this time of gypsum formation; it was between 80,000 and 40,000 years ago. Once again there was uplift and the gypsum surface west of Lake Eyre was eroded whilst the lake bottom was attacked by the wind.

Some of the material deflated from the lake accumulated in dunes to the north. There is some evidence to think that this wind action chiefly took place between 40,000 and 20,000 years ago. However there have been alternations of water erosion and deposition and of wind erosion and deposition down to the present time. During this final period, salts have been precipitated from the streams flowing to this centre of internal drainage to form the present salt plain of Lake Eyre.

The Volcanoes of Australia

Australia is the only great land mass in the world at the present day entirely free from the violence of volcanic eruptions. This has not always been so. From the past, as in every other land, comes the story of a succession of such eruptions, of destructive floods of lava, showers of incandescent ash, hot springs and geysers, and the accompaniment to all these, violent earthquakes.

Some of the volcanoes have left evidence of their passing in the present contours of hills and valleys, others are responsible for spectacular scenery. More often, and this applies particularly to those of the earlier periods, the shapes of the volcanoes have disappeared, and all that is left is richer and more productive country. The farmer speaks of rich volcanic soil, but few who speak thus pause to visualize the scene as it must have been when the volcanoes were active.

Volcanoes, though they have been active at some time in all parts of the world, and at all times in some part of the world, are actually one of the minor manifestations of the power of nature. They are really a result rather than a cause. The popular conception that the earth is riven into great gulfs and that mountains are upheaved by the action of volcanoes and earthquakes should be reversed. It is the volcanoes and earthquakes that are the result of the earth movements which raise mountains and depress valleys and plains beneath the sea.

If the earth can be imagined as a plastic body, intensely hot in the centre, but kept solid by terrific pressure, any release of pressure will allow the underlying material to become molten at once. Such release of pressure takes place when there are movements of the earth's crust, particularly those lateral movements by which alpine ranges are crumpled up. Without going into the causes of such lateral movements, their effects may be seen in many places where rocks are exposed to view. When a pile of paper is squeezed from the sides it rises in the centre, and in the same way rock strata when compressed by great lateral earth thrusts are crumpled into folds. The folds may be simple or they may be very complicated, and the rocks are often fractured and shattered so that deep cracks or lines of weakness develop. Beneath the folds pressure becomes reduced, and the deep-seated material, already at a high temperature, melts and forces its way upwards through vertical cracks to the surface as lava. The eruptions are often accompanied by explosions caused by the expansion of incandescent gases, and the surrounding rocks are further shattered and hurled into the air as volcanic ash. The coarser part of the ash falls close to the vent and builds up volcanic cones, the finer ash is carried by wind far and wide, and may cover the surrounding country or the bottom of adjacent seas to a considerable depth.

Earthquakes are also a result of earth movements rather than their cause. They are the vibrations set up when rocks are fractured or displaced by any disturbance of the earth's crust. The vibrations spread outwards from the centre of disruption; although their intensity diminishes with distance, they may be recorded on delicate instruments thousands of miles from their source. Australia has long been free from major earthquakes, but even in its present state of comparative stability small earthquakes are occasionally felt. These are due to minor movements of the rocks, subsidence in such areas as Lake George in New South Wales or St Vincent Gulf in South Australia. It is extremely unlikely that Australia in the immediate future will suffer any of those terrible catastrophes which from time to time devastate some other countries. However in October 1968 a rather stronger earthquake than is usual in this country opened fissures and dislocated the ground several feet along a north-west line running through Meckering in Western Australia some 80 miles from the Darling Range and fault. Much damage was done to buildings in the town but there was no loss of life.

The earliest known volcanic eruptions in Australia occurred at the dawn of geological time, in ages long before the beginning of the Cambrian Period. Some were on a gigantic scale and were spread over a great area. It is not known where the original craters were centred, but the lava-flows were of great extent. In the Kimberleys they aggregated over 3,000 feet in thickness. Lava flows of the same age are found in the Mt Isa ranges and there was similar activity in the Flinders Ranges in South Australia, not of the same age but still belonging to the Proterozoic. Then later on in the Cambrian Period the very extensive basalt flows of the Antrim Plateau in the Kimberleys which have already been mentioned spread out over the land. Much of the old lava has long since decomposed and been removed by erosion, and only in certain areas has it any effect on the present landscape, except when rich patches of soil mark its position beneath the surface.

Though the very early volcanic eruptions are now barely discernible in the landscape, their early cessation has given particular character to the western part of Australia. After their termination the centre of the scene moved, and henceforward volcanic activity was practically confined to the extreme east. One small exception occurs in the Fitzroy valley in West Kimberley where there are a number of volcanic plugs of a rather special dark igneous rock called lamproite, which are of Miocene age. Most of them form prominent, if low, hills. Western Australia is thus very nearly unique. It is doubtful if there is elsewhere any region which has for so long been as stable and free from major disturbances.

The story is very different on the eastern half of the continent. In each of the early periods succeeding the Proterozoic Era there were phases of intense volcanic activity, now in this part, now in that. There is here no need to give a chronological account of them, except where they have given a special feature to the present landscape. Their exact age is less important to us than the composition and hardness of the lavas and other rocks erupted, the effect upon them of subsequent movements of the earth's crust and their evolution into the present landscape.

Many volcanic rocks are hard and resist erosion better than such sedimentary rocks as shales and sandstones. Others decompose rather rapidly when exposed to the air. Dark-coloured lavas such as basalt often decompose rapidly to pure clay, and as they are rich in iron they form deep red soils. Other light-coloured lavas such as trachytes and rhyolites are rich in silica, and may be exceedingly resistant to both erosion and chemical decomposition. So also are other rocks such as the various types of porphyry. These were not poured out as lavas but were also originally in a molten condition. In this state they failed to reach the surface, but after forcing their way upwards as intrusions through other rocks they cooled slowly within the earth. Or they may have filled the vents of old volcanoes, cooling slowly in the final stages of the eruption.

Owing to its hardness this type of rock often remains long after the surrounding softer rocks are worn away. Its remnants form the tops of hills, or rugged cliffs on the sea shore, or stand as giant monoliths above the level of the plains.

The great rocks at the entrance to Port Stephens in New South Wales were formed in this way. Either from the sea or from the land they present a remarkable sight. The surrounding country is flat, much of it swamp a little above sea level, and from it the monoliths rise almost sheer to a considerable height. The highest are on either side of the entrance to Port Stephens, the northern one over 700 feet, the southern one over 600 feet high, and they form magnificent gateways to a noble port.

Similar monoliths lie for some distance inland, and give character to much of this area, northwards to Taree and westwards to "the Buckets", curious mountains overlooking the little township of Gloucester. Not all are intrusions of porphyry, for many consist of the lava rhyolite, poured out during the Carboniferous Period. Most of the sandstone has disappeared in later ages, but the porphyry has defied destruction, and now forms most of the higher peaks throughout the district.

Somewhat similar are the peculiar Glass House Mountains in southern Queensland. Their discovery goes back very early in Australian history. When the *Endeavour* was in Moreton Bay in 1770 Captain Cook's attention was attracted by the appearance of a number of great isolated rocks rising above the coastal plains. A passing shower had moistened their summits so that they glistened like glass in the distance. He thereupon named them the Glass House Mountains, and the name has been retained to the present day.

The rocks are only about 50 miles from Brisbane. The railway here runs northward over a low sandy plain, skirting them on the eastern side for some

15 miles. All are within three or four miles of the railway, and some are clearly visible from the train. The country is low and undulating, very little above the level of the sea, the soil poor and sandy with many patches of swamp. A straggling forest of blue and white gums, of stringybark, teatree, banksia and casuarina fills the landscape for many miles. There are also extensive plantations of pine. Beneath the trees the view is restricted to a few yards, hence when from some open space one of these strange mountains suddenly looms into sight its magnitude seems stupendous by sheer contrast.

Every two or three miles one of them rises sheer from the plain, some isolated and apart, others grouped, particularly towards the northern end. Most are bare rock to the summit, but on the tops of some a few stunted eucalypts have taken root. The majority are under 1,000 feet in height, but Mt Tunbubudla is 1,106 feet, Mt Coonowrin 1,231 feet, and Mt Beerwah rises as a sharp pointed cone 1,819 feet. Perhaps the most striking of all is Mt Coonowrin, shaped like a cylinder with a truncated top and smooth vertical sides.

The Glass House Mountains, according to the noted geologist, Dr H. I. Jensen,[7] are the eroded remains of true volcanoes. The eruptions occurred in early Miocene times. Potassium-argon dating has shown them to have formed about 24 to 25 million years ago. The rocks underlying the plains in this area are the Ipswich Coal Measures, and these were shattered and bent by the molten rock as it forced its way upwards. The molten matter later cooled into the light-coloured rocks known as rhyolite and trachyte. These rocks are more refractory than the darker basalts, and seldom flowed out as lava flows upon the surface. Instead as they hardened rapidly, they plugged the vents, then to be shattered by tremendous explosions, fragments accumulating as volcanic ash upon the sides of the growing volcano. In the last stages of eruption the plugs of lava within the vents would remain intact, solid cores of hard rock within cones of more friable material. Now the outer portions of the volcanoes are gone, most of the ash long since worn and washed away, and only the central cores of rhyolite and trachyte still remain, lone monuments in a lowly landscape.

Similar in many ways to the Glass House Mountains are the Warrumbungles and the Nandewar Range in north central New South Wales. Both are prominent features in the present topography.

By the Eocene Epoch the whole eastern part of Australia had been very nearly worn to one vast plain. It was from this time that an upward movement of the land began, to culminate in the formation of the tableland flanking the eastern margin of the continent. The interior of the country was unaffected, and the tableland on its western side generally slopes downwards to the great western plains. Along the whole belt of the central western slopes there are few land features of any magnitude; hence the few mountains which do exist seem magnified by comparison with their surroundings.

The Warrumbungles, impressive in themselves, are made more so for this

[7]*Proceedings Linnean Society, New South Wales,* 1903.

reason. Though not of outstanding height they are visible for a great distance. The township of Dubbo lies about 60 miles to the south, just where the lower slopes of the tableland begin to merge into the western plains. From the hills about Dubbo on a clear day, particularly when the setting sun magnifies the distant view, small irregular hummocks appear indistinctly above the northern horizon, so far away that even a powerful glass gives little clue to their size and extent. From a somewhat nearer view they appear as a number of isolated peaks. On approach more and more are seen above the horizon, until at last all are merged into a rugged mass of mountains rising from 1,500 to 3,000 feet above the general level of the country. The jagged sky-line of the mountains is clean cut in the clear inland air, rising in conical peaks or standing out as vertical needle-like monoliths above the grey green of the wooded heights. Here and there are the lines of deep gorges penetrating the mountains right to the centre.

The whole occupies an area roughly circular and over 30 miles in diameter. On the eastern side, nestling in a valley half enfolded by outlying spurs, lies the little township of Coonabarabran, on the banks of the upper reaches of the Castlereagh River. This river rises in the deep heart of the Warrumbungles, flowing at first to the east, then describing a complete half-circle about the southern flank of the mountains until it turns westwards on its course to the Darling. All the streams radiating from the mountains are tributaries of the Castlereagh, and like the main river are at first perennial in their flow. Away from the mountains they either evaporate or sink into the ground, and in normal seasons become mere chains of waterholes or disappear altogether. After heavy rains they become considerable rivers, often overflowing their banks to merge into an inland sea many miles in width.

To the north of Coonabarabran the Warrumbungles continue as a ridge a few hundred feet high forming the divide between the Castlereagh and the Namoi. North of this again is the Pilliga Scrub, a vast expanse of rather poor sandy soil, with low mesas of sandstone, all covered with a dense forest of cypress pines.

The Pilliga Scrub, though now partially opened for settlement, is still an area where the newcomer is easily bushed. Here is a useful tip for anyone who may travel across country and away from the roads. Of the two species of cypress in the area, the bark of the white pine (*Callitris glauca*) alone invariably supports a rich growth of moss and lichen. This is beautifully oriented, always growing on the south side, and so clearly defined that by taking a median line true north may be found within one or two degrees.

The Warrumbungle peaks are the remnants of true volcanoes, and in age are a little younger than the Glass House Mountains because they erupted about 15 million years ago in the Middle Miocene. It is hard to say just why eruptions on such a scale should have occurred in this particular region. The underlying rocks here are sandstones laid down in a gigantic lake which filled the centre of Australia in Jurassic times, that is about 150 million years ago. The sandstones have lain virtually undisturbed ever since, and beyond a slight uplift there have been no major disturbances of the earth's crust. In folded mountain ranges such as the Rockies and the Andes volcanic eruptions follow as a natural corollary; here they came from no ascertainable cause, shattering the stillness of a landscape which had lain peacefully for so many long ages.

It is not known how many individual volcanoes there were. There must have been many large as well as innumerable smaller centres of eruption. The main craters have long since disappeared, but everywhere is the evidence that they once existed. The evidence is mainly the solid vertical cores of lava which had solidified within the vents and now remain after much of the overlying and surrounding beds of ash have disappeared. The story of what happened is fairly clear.

Prior to the first eruption the Jurassic sandstone had been elevated into a low tableland which later became considerably higher. At Coonabarabran this is now 1,700 feet above sea level. On the western border of the Warrumbungles it is 1,200 feet high. The climate was moister than it is now. The rivers had a greater flow and excavated their beds to some depth. When the first eruptions came there were violent explosions, and from the débris numerous volcanic cones were built up. Next came great lava-flows which not only covered the original cones but spread far and wide, burying the neigbouring river valleys to a great depth. Here and there in the bottoms of the valleys leaves and other vegetable matter were buried in the clay beneath the lava and preserved. Many of the plants of this time were typical of a warmer and moister climate, and the fossil leaves are indeed similar to many living now in the rain forests on the coast.

The lava-flows were viscous and did not travel far, and as they merged together they became heaped higher and higher until they formed a mass of volcanoes thirty miles across and probably much higher than they are now. Most of the lava in the early stages was trachyte, a light-coloured rock similar to that of the Glass House Mountains. Last of all came flows of dark basalt, much more fluid than the trachyte; some of these flowed much farther and spread beyond the perimeter of the main mountain mass.

Geysers or hot springs were active at various stages. Some formed lakes of some size, the water of which was hot or at least warm. Very few organisms can live in such lakes, but there was one well able to adapt itself to the conditions, just as it does today. This was a microscopic unicellular plant of very lowly organism, one of the order of diatoms, plants secreting very beautiful siliceous shells of complex design. Diatoms play a very important part in the scheme of life. They live almost anywhere there is water, whether it be hot or cold, fresh or salt. They occur in countless millions in all the seas of the world, from the tropics to the polar regions, and they form the basic food of most of the animals in the ocean. They live also in the warm springs and lakes of volcanic regions, and in the Warrumbungles as elsewhere their remains have accumulated to form deposits many feet in thickness. Such deposits are of the white friable rock known as diatomaceous earth or by the trade term, fuller's earth.

When the eruptions ceased at some time in the Tertiary Period the mountains must have been much higher than they now are. During the many millions of years that have since elapsed they have suffered from the ravages of time. During periods of more effective precipitation in the Pleistocene there would be more runoff and the rivers would have been capable of more erosion than at present. All traces of the original craters have disappeared, much of the volcanic ash has been washed away, and only the hard plugs of lava originally filling the vents remain as vertical pillars towering high into the air.

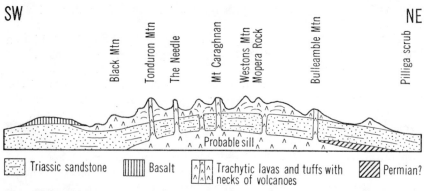

11. Section through the Warrumbungles (after Jensen) showing the main centres of volcanic eruption.

Most of the higher peaks are composed of such pillars or are capped by the remnants of the harder lava-flows. Two of them, Mopera Rock and Westons Mountain, are each more than 3,500 feet high, and stand like sentinels on either side of the gorge in which the infant stream of the Castle-reagh first sees the light. About seven miles to the west of these is the highest point of all, Mt Wambelong, 3,954 feet, but more spectacular is the lower Crater Bluff, a giant monolith, its topmost thousand feet a nearly vertical rock.

About 60 miles to the north-east of the Warrumbungles area is another almost identical group of old volcanoes, the Nandewar Range. The peaks begin to rise on the northern side of the wide flat valley of the Namoi River and occupy an oval area more than 20 miles in length. Seen from afar they appear as a domeshaped mass with a jagged skyline, averaging over 4,000 feet, with many peaks approximately 5,000 feet above sea level. A nearer view shows them as a massive wall hiding the higher peaks behind; nearer still many spurs are seen leading down to the plain, while between them narrow gorges soon wind out of sight.

On 18th December 1831, Sir Thomas Mitchell, seeking a way to the north, wrote in his diary: "I rode forward with Mr White and the native and soon entered an extensive valley, beyond which I could just perceive, through the general smoke, a majestic chain of mountains extending to the westward, the

Wilpena Pound, Flinders Ranges, South Australia. The basin occupies a downfold in quartzites. *South Australian Tourist Department.*

The Breadknife, Warrumbungle Mountains, New South Wales. A dyke exposed by erosion. *New South Wales Department of Tourism.*

Tonduron Spire, Warrumbungle Mountains, New South Wales. Core of a Tertiary volcano isolated by erosion. *J. N. Jennings, Australian National University.*

rocky masses terminating in pointed obelisks, or broken into bold terraces of dismal aspect." Mitchell's appreciation of the scenery was no doubt modified by the fact that the mountains lay right athwart his intended route, and to his clumsy cart transport presented a seemingly impassable barrier. Nevertheless he sought a way across, and a day or so later ascended a high peak for this purpose. He reached the summit only to find further progress blocked. "The only link connecting with hills still higher was a very bold, naked rock, presenting a perpendicular side at least 200 feet in height. To proceed further in that direction was quite out of the question."

The mountains today present exactly the same aspect as they did to Sir Thomas Mitchell over a century ago, except that the plains are now intersected with roads and dotted with farms and sheep stations, and the railway runs through the town of Narrabri quite close to the nearest slopes. But the mountains themselves are still virgin and untamed, sheer cliffs rising abruptly above the narrow gorges, and steep slopes covered with straggling forest reaching up to the lofty pinnacles above.

One of the best vantage points for a close view is the valley of Bullawa Creek, which penetrates the heart of the mountains, and is overlooked by some of the highest peaks. Among these is Ningadhun, 3,602 feet, which rises in a vertical pillar of rock above a regular cone. The highest peak of all is Kaputar, 4,949 feet; others not much lower are the twin peaks, Mt Lindesay, 4,505 feet, and Round Mountain, 4,396 feet, and there are many more of equal or greater height in this vicinity.

The story of the Nandewars is that of the Warrumbungles. There were initial explosions and showers of ash to build up cones, then flows of trachyte to form the main mass of the mountains. In the final stage there were many minor eruptions of basalt from small volcanoes well beyond the perimeter of the main eruptions. Finally came the lowering and partial destruction of the mountains by the slow process of erosion. No doubt the Nandewars were originally much higher than they are now, but their summits are still amongst the highest in the whole of Australia.

Perhaps greater than any of the volcanoes so far described was the huge Tweed volcano, the rims of which straddle the New South Wales-Queensland border in the Eastern Highlands. Mt Warning and its neighbouring hills are the intrusive plugs of this volcanic pile. They form a central relief surrounded, beyond an annular depression, on all sides except seawards by a rim of basalt ranges with steep, inward-facing scarps, the Macpherson, Tweed and Nightcap Ranges. The basalts decline much more gently away from this centre, representing the lower, outward flanks of a huge shield volcano, which reached majestically many thousands of feet above the neighbouring parts of the Eastern Highlands. It was active in early Miocene times about 25 to 22 million years ago.

The Glass House Mountains, the Tweed volcano, the Nandewars, the Warrumbungles and the Canobolas overlooking Orange farther to the southeast all form a belt of former volcanic activity, the cause of which needs much explanation. Curiously enough, this belt resembles others which existed about the same time in other parts of the world, and differs considerably from the present active volcanoes. The latter for the most part are associated with regions of intense folding, or are connected with the warping of the earth on the margins of deep seas. The lavas of the early Tertiary volcanoes are also nearly always trachyte or an allied rock, and are so rich in potash and soda that they are classified under the general term alkaline rocks. To trace the origin of the Australian volcanoes it is necessary to go back much further in time. Many lie on the edge of the Great Artesian Basin of the interior, and all are extruded through sandstones which dip far below the surface to the westward.

The Artesian Basin, on which so much of our prosperity depends, is a vast depressed area covering much of western New South Wales and Queensland, and extending well into South Australia and the Northern Territory. This area sank intermittently through three whole geological periods, the Triassic, the Jurassic and the Cretaceous, in all for over 100 million years. For a long time in the Cretaceous it was so low that the sea came in from the north and divided Australia into two islands. Indeed for a short time it probably divided the continent into three portions. Before and after this the land was covered by vast lakes, the earlier lake in Triassic and Jurassic times covering no less than 300,000 square miles. In these lakes and in the intervening sea rocks more than a mile in thickness were deposited, sandstone for the most part, but also shales and limestone. The enormous weight of the accumulated sediments removed from the high land at the sides and piled in the centre altered the distribution of pressure in the plastic mass below the earth's crust, causing the margins to rise while the centre continued to sink. This no doubt was one of the factors which tilted the eastern margin of the continent into the present tableland.

Pressure was lessened beneath the uplifted portion and the deep-seated material became molten. Fractures and lines of weakness developed on the margins of the main basin, and through these the molten material forced its way upwards to erupt on the surface as volcanoes. It was a gradually diminishing process, tending always to restoration of stability, until at last equilibrium was reached, the volcanoes became extinct, and only their remnants now remain as witness to their former fury.

From volcanoes such as the Nandewars and the Warrumbungles comes much that is spectacular in scenery, but before and after them and probably coeval with them were other eruptions in eastern Australia on a still greater scale, even if their effect on the present landscape is less apparent.

These eruptions began very early in the Tertiary Period, at least 58 million years ago, and recurred right through that period, the Pleistocene and even into the Recent. Although they built up no large volcanoes or high mountains, they nevertheless played a big part in fashioning the shape and character of the whole east coast from Cape York to Victoria.

Most eruptions were of the fissure type. This means that there were no well defined centres of eruption, no great craters at the summit of high volcanoes; the lava merely welled forth from extensive fissures in the earth's crust. Many of the fissures may well have been caused by the uplifting of the eastern plateau in the Tertiary Period.

All the lava-flows were of basalt, which melts at a comparatively low temperature and thus retains its fluidity for considerable distances. In most districts two major phases of activity can be recognized, for newer lavas have flowed into and filled valleys produced by erosion in the surfaces of much older ones. The terms "older" and "newer" basalts are generally used by geologists in all the States, and this classification is on the whole sufficient and satisfactory. The "older" basalts span most of the time from the beginning of the Tertiary whereas the newer basalts which preserve much of the detail of the landforms they took on when they solidified belong to the Pliocene, Pleistocene and Recent.

The floods of lava covered vast areas of country. They filled the river valleys. They both heightened and levelled the summits of tablelands and poured out over the plains. The oldest of the flows have naturally suffered most from erosion. In the Macpherson Ranges, on the Blue Mountains, in the Monaro, in the high lands of Victoria, they have been completely cut through by the rivers, and often remain as residuals on the summits of the hills. Beneath them the beds of former streams are often buried, the old alluvium rich in gold, tin and precious stones. These are the famous deep leads of the miner from which much mineral wealth has been won.

The surface of the newer flows may be so little eroded that it remains now much as when it was poured out. The surface rock is often decomposed to form patches of rich soil, and these when cleared make the most productive dairying country in eastern Australia.

Another feature of the basalt-flows is their close association with the rain forests of the east coast. Basaltic soil is particularly favourable to the growth of the jungle, and where the climate is warm and the rainfall high there the jungle flourishes. During the climatic oscillations of the Pleistocene and the Recent, the jungle no doubt advanced from and retreated to the coastlands. Certainly it advanced westwards in the Atherton Tableland between 10,000 and 7,000 years ago, probably as a result of increasing warmth. At the coming of the white man it covered great areas on the coast and tableland. Now much of it has disappeared before the axe and the plough.

The newer basalt-flows attained their maximum in northern Queensland. The tablelands from north of Cairns to the south of Townsville are nearly all covered by basaltic lavas. Between the Herbert and Burdekin rivers they cover over 2,000 square miles at an average height above the sea of 2,000 feet. Many of the flows are quite fresh, and one can be traced for over 25 miles. Rising in many places are intact cones, the centres of former eruptions. Mt Fox is a perfect example of an extinct volcano. Its summit, 2,662 feet above sea level, is a truncated cone, with an area of two or three acres, the depression forming the crater surrounded by a rim about 14 feet high which has been breached on one side by the flowing lava. No trees grow on the sides of the cone as the lava is still too fresh to have decomposed to soil. This volcano must have been active in recent times.

Another series of later lavas extends west of Charters Towers for over 150 miles, and forms a plateau 2,500 feet above sea level. Numerous centres of eruption are still visible, and Mt Emu is a well-preserved crater of the explosion type.

The Atherton Tableland was formerly densely covered with jungle, much of which has been cleared for farm lands. Within one of those few areas where it is still virgin are Lake Eachem and Lake Barrine, clear blue deep bodies of water, with the dense jungle coming right to the edge of the water. Both lakes are surrounded by ridges of coarse volcanic ash, and it is evident that they occupy former craters caused by tremendous explosions in the concluding stages of volcanic activity.

Farther south the valleys of existing rivers are partially filled by recent lava-flows, the Musgrave River and the head of the Endeavour River among others. In the valley of the Burnett River the flows are so fresh that they still retain their slaggy surface, and here also is the perfect cone and crater of Mt Le Brun.

In southern Queensland and northern New South Wales the basalt-flows belong to the older periods; in New South Wales it is doubtful if there was any volcanic activitiy after the Tertiary Period. The earlier flows were of great extent, and there are many basaltic plains and tablelands formerly covered with jungle which are now rich farm lands. Such were the Kin Kin Scrub near Tewantin in Queensland, the Great Scrub formerly covering the Big Bend of the Richmond River in northern New South Wales, the Dorrigo Scrub, the Lansdowne Scrub and smaller areas.

On the south coast of New South Wales there are patches of basalt which belong to a much older period. These flows were actually poured out in the Permian Period, and are now only visible on the surface because great thicknesses of overlying rocks have been removed by erosion. From Wollongong to beyond Kiama they form the coastal plain, and abutting on the sea, have been worn by the waves into a rugged coast, relieved by the long stretches of beach.

At Kiama the basalts are columnar, consisting of hexagonal columns packed closely together at right angles to the surfaces of the original flows. This structure is a curiously regular one, produced evidently by an even contraction of the rock when cooling from a molten state.

The topography of the coastal plain is typical of basalt country, smooth rounded hills and flats of rich red soil, formerly thickly forested, but now grass-covered and used for pasture. Innumerable small loose rocks, fragments which have resisted decomposition, have been gathered and built into low walls between the fields, replacing the wooden and wire fences in use elsewhere. There are even traces of the jungle, for this is about the last outlier of the tropical forests to the north. In sheltered corners are odd jungle trees, particularly the broad-leaved stinging trees, its innocent-looking leaves still waiting to entrap the unwary passer-by.

In eastern Victoria there is a great development of the earlier basalts, particularly in the southern portion of the high tableland. Most of them were poured out prior to the uplift of the tableland. They have been deeply dissected by erosion, and they are intricately mixed with many other rock formations to shape the present topography. Some of them are very hard, and where they overlie soft Ordovician slates, their remnants now cap some of the highest peaks. The youngest of these older basalts of eastern Victoria has been given the date of 17 million years but the oldest goes back to the very beginning of Tertiary time in the Paleocene Epoch.

It is in western Victoria that there is abundant evidence of the most recent volcanic activity in Australia. Here the Great Western Plain has many points in common with the jungle-clad Atherton Tableland in Queensland, but scenically it is in sharp contrast. This is an extraordinary region, a vast area of country a little under 500 feet above sea level, almost a perfect plain, but relieved by hundreds of small isolated hills and 27 lakes.

The length of the plain is over 160 miles, reaching from the Glenelg River in the west to Port Phillip in the east, and its maximum width is about 75 miles. In total area is about 9,000 square miles, and it is largely composed of lava and volcanic ash to a depth of hundreds of feet. Of its kind, that is, a plain formed entirely by volcanic action, it is the third largest in the world, exceeded only by the Snake Plain in the United States and by the Deccan in India.

It has another distinction. With Mt Gambier, just across the border in South Australia, it shares with parts of Queensland in being the last part of Australia to suffer the fury of volcanic action. The oldest lavas have been dated by the potassium-argon method as 4,500,000 years old, that is late Pliocene. The youngest craters to be dated look extremely fresh indeed. Radiocarbon dating of charcoal, shells and bones above and below the ashes erupted from Tower Hill near Warrnambool in Victoria show that the volcanic activity occurred between 5,500 and 6,000 years ago. Charcoals beneath ash layers of Mt. Gambier across the border have also been dated in this way and tell of two eruptions, 4,800 and 1,400 years ago, from this volcano, the youngest of all in this country.

Scenery in the region is not on a grand scale. There are no rugged mountains, no deep gorges, no dense forests nor waterfalls. The plain instead is sparsely timbered, and its fertile expanse is dotted with prosperous farms and orchards and thriving townships.

One feature of the country is the absence of visible rivers and creeks. Some do exist, and Mount Emu Creek traverses the whole plain from north to south, to join the Hopkins River and reach the sea at Warrnambool. Even this stream is invisible from a short distance, for it has already cut its channel to a shallow gorge about 20 feet below the level of the plain. The absence of surface drainage is probably accounted for by the porous volcanic ash underlying so much of the country. Much of the rain seeps underground to form a sub-artesian basin and a perpetual reservoir of good water for the farmer. Every farm in the district has its wells and windmills to pump the water to the surface.

The underground seepage probably accounts also for the water in the many lakes. Few of the lakes have a sufficient catchment area to provide the deep volume of water within them. It has been noticed in confirmation of this that the level in some lakes alters with the artesian watertable following extended periods of good or scanty rainfall. The changing water levels in Lake Keilambete over the last 35,000 years have been reconstructed from the shorelines and sediments left behind. Radiocarbon dating provides the chronology. There have been at least five oscillations from high level to low and back again in this period, but on differing scales. These changes are certainly partly due to temperature changes causing variations in evaporation as well as to precipitation changes so it is not easy to say exactly what climatic changes are implied by these variations in lake level.

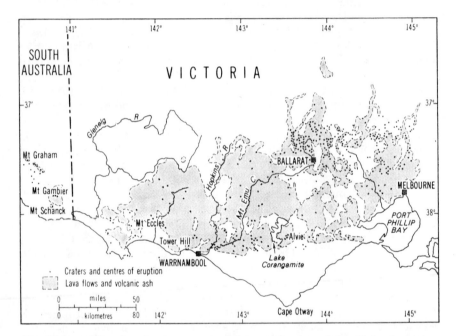

12. Western Victoria, showing extent of latest volcanic activity. Volcanic centres from unpublished map by E. B. Joyce and lava flows from Geological Survey of Victoria.

From any of the hills the view is wide, but from Mt Noorat, almost centrally situated, it covers practically the whole extent of the plain.[8] Mt Noorat is an extinct volcano with a beautifully preserved crater, about 440 yards in diameter, nearly circular, and surrounded by a very narrow rim. It is a mountain by comparison only, rising about 500 feet above the general level. It lies about four miles north of the township of Terang. From its

[8] H. J. Grayson and D. J. Malony, *Memoirs, Geological Survey, Victoria,* ix, 1910.

summit and far to the north can be seen the shadowy summits of the Grampians, marking the northern limit of the plain. Far to the south again are the Otway Ranges, and adjacent to them the line of forest separating the plain from the sea. To the east, about 12 miles away, is Bullenmerri, a prominent gently-sloping dome, on the side of which the town of Camperdown is built. Within Bullenmerri are the twin crater lakes Bullenmerri, which is fresh, and Gnotok, which is salt, but from this viewpoint both are invisible. One lake is visible, Keilambete, its vivid blue water lying in an almost circular basin nearly a mile in diameter. From this distance its low rim, gently sloping outwards but with a steep inner escarpment, is barely discernible above the plain.

All the hills within sight were centres of eruption, and many have the original craters in a perfect state of preservation. One of the highest and most conspicuous is Mt Elephant, 12 miles to the north-east, so called from its fancied resemblance to a kneeling elephant. This is steep-sided and symmetrical, with a single huge crater, breached on one side by an old lava-flow. Mt Elephant is 1,290 feet above sea level and 780 feet above the plain. Of the other hills Mt Leura has had the main crater partly destroyed by an explosion, and a secondary cone has been built up, partly on the rim and partly within the old crater.

Most of the volcanoes are scoria cones, that is, they were built up of layers of volcanic ash ejected from the central vent. Others are of solid lava which welled up quietly and solidified in a circular ring. Mt Hamilton in the north is an example of a lava cone. Then there are composite volcanoes built up of alternate layers of lava and ash. Amongst them are Mt Rouse and Mt Napier in the north-west, Mt Eccles near Macarthur, and a number of centres near Portland such as Mt Clay and Mt Eckersley. Here the crater is deeper than the surrounding plain.

Amongst the most remarkable of the lakes are Lakes Bullenmerri, about two miles across, and Lake Gnotok, which is quite close and nearly as big. Both lie in original craters caused by tremendous explosions, but while their bottoms are practically on the same level there is a great difference in the depth of water. Bullenmerri is 263 feet deep, and the salt Gnotok only 103 feet, so that the surface of the former is much higher than that of the latter, in spite of their being so close to each other.

The lakes are true crater lakes, occupying the cavities caused by sudden terrific explosions. However some circular volcanic depressions of large size are due to subsidence. This subsidence is caused by collapse of parts of volcanoes into the space created by the flowing and blasting out of lava from them. In western Victoria the only features which belong to this category of subsidence calderas, as they are called, are Bridgewater, Nelson and Giant Bays west of Portland. Here more than half of each caldera has been lost to the sea through wave attack.

Many of the explosion craters of the Western District are of a special nature because ground water in the limestone beneath the plains reached the hot volcanic vents and caused short-lived explosive activity, which shattered both lava and bedrock. These fragments were thrown out to form low rings of mingled ash and sedimentary rock débris. Many of the volcanoes here are tuff rings of this type, some with lakes in them, some without. Such are Tower Hill, Mt Leura, Lakes Bullenmerri, Gnotok, Purrumbeete and Terang. At Tower Hill and Mt Leura younger scoria cones were built up by later activity within the rings.

Just across the border in South Australia lies the westernmost outpost of recent volcanic activity in Australia. Coeval with and of the same type as those in western Victoria, the extinct volcanoes adjacent to Mt Gambier form a small group by themselves. Right at the close of volcanic activity a fissure appears to have developed from the border of the sea northwards for more than 20 miles, and along this fissure a number of volcanoes erupted streams of basaltic lava and spread showers of volcanic ash over the surrounding country. Eight main volcanoes can be recognized from Mt. Schank in the south to Mt Graham in the north. The largest and most beautiful of them it Mt Gambier, the perfect rim of its wide crater rising about 600 feet above the surrounding plain. The beds of volcanic ash of which the rim is composed still lie at an angle parallel to the slope of the hill, and are as fresh and undecomposed as when they were erupted.

Within the main crater of Mt Gambier nestle a whole series of small lakes, the beautiful Blue Lake, which supplies the township of Mt Gambier with water, Browne Lake, Valley Lake and Leg of Mutton Lake. The last occupies the crater of a secondary cone within the main crater. Deepest of all the craters is that of Mt Schank, its sides, like those of Mt Gambier, composed of volcanic ash. The inner slope of the crater is very steep and goes down for some hundreds of feet below the level of the outside plain.

When the last volcano in this region became extinct, all volcanic activity ceased in Australia. For a long period great areas had been rendered desolate and barren; neither vegetable nor animal life could have lived on the slow-cooling fields of black lava or beneath the continuous showers of incandescent volcanic ash. The skeletal remains of aborigines 8,000 years old have been found in Western Victoria and tools they fashioned have been found beneath ash from Tower Hill at Koroit Beach. Even though there is no mention in his folklore and legends, the Australian aboriginal must have watched fearfully the terrifying spectacle of this violent aspect of natural forces.

One other fact is worthy of thought. What nature takes away with one hand she is apt to restore with the other. From devastation comes ultimate fertility. Basalt is the lava richest in those ingredients which provide nutriment for living things. Short though the time since volcanic eruptions ceased, the surface of the lava and ash has already decomposed sufficiently to cover the land with a good depth of the richest soil, and to convert a place of desolation into one of the most fertile in Australia.

Crater lakes at Alvie, Western District, Victoria. Lake Corangamite in rear. *Victorian Railways.*

Port Jackson. The other drowned valleys of Middle Harbour and Broken Bay are in the background. *The New South Wales Department of Tourism.*

Wave-built sand barrier separating Smith Lake from the sea, North Coast, New South Wales. The lake occupies a drowned valley. *J. N. Jennings, Australian National University.*

Grose River Valley, Blue Mountains, New South Wales. The river has dissected the sandstone plateau to reveal underlying shales. *New South Wales Department of Tourism.*

Port Jackson

When the First Fleet sailed through the Heads into Port Jackson over 180 years ago Governor Phillip was astounded at the spacious waterway stretching before him. Lofty sandstone cliffs a little more than a mile apart, formed a gateway. Directly opposite were more sandstone cliffs forming another gateway to the long winding reaches of Middle Harbour; while to the left the main harbour bent once and again to the right, mile upon mile, to the silent waters of the Parramatta River. Forests of gum trees crowned the hill-tops and the slopes reaching down to the water's edge. Here and there a few rocky islets rose above the surface. Natives appeared in groups upon the shore, half terrified, half fascinated by the great bird-like creatures swimming upon the sea. Here was an untroubled virgin land, the last of all the continents save Antarctica to come within the influence of the white man.

Through the same Heads now come great liners, very different from the tiny sailing ships which pioneered the way. Here, from a hundred ports, come cargo vessels and at times the warships of our own and allied nations. On the tree-girt hills a great city has grown apace, one of the great cities of the southern hemisphere. A lofty bridge, emblem of man's constructive ability, spans the waterway. Buildings tower skywards to look over streets crowded with traffic; there is the hum of industry and thousands of homes crowd every available space.

It would appear a great and lasting change, but it is only a veneer upon the fundamental structures of nature. In the long story of the land it is the city that is the superficial thing, making little appreciable impression on the course of natural events. In the infinity of time the works of man are the most transient of all.

The history of Port Jackson began a long time ago. The close of one of the great geological periods, the Permian, may be taken as a starting point, and this was 230 million years before man was on the earth. In the following period, the Triassic, the foundations of today's coastal geography were laid. The picture of events in Triassic times is fairly clear, as is also the appearance of the old landscape. The country to the north and south of Sydney was flat, as was the area immediately to the west. It was little above sea level, and had lain so since coal seams had been formed in the Permian swamps.

The Blue Mountains were not to be born for long ages to come, but a range of mountains ran westwards through Cobar to Broken Hill, and there were high mountains over New England and the Monaro. The sea was farther to the east than it is now. A great gulf opened to the south, separating other lands far to the east, lands joined to the Australian continent in the vicinity of Brisbane. This was the head of the gulf.

Then the land began to sink. It sank slowly in the form of a basin for a long age, low enough to allow extensive freshwater lakes and swamps to form, and with the eastern rim high enough to exclude the sea. Rivers flowed into the lakes and swamps, depositing sand and mud, burying plants which grew there, as well as leaves and branches washed from the neighbouring land. Strange fish lived in the water, and their bodies too have been buried and preserved. Even the bodies and wings of primitive and extinct insects were permanently entombed, and more rarely the skeletons of the curious frog-like creatures called labyrinthodonts. These were amphibious in habit and sometimes as large as a horse. Footprints have also been found in the sandstone, probably of some large strange reptile, the form of which can only be surmised.

All of this is written in the rocks in the brick pits at St Peters, Willoughby and Brookvale, and in the cliffs at Narrabeen and elsewhere. Even a passing shower has left imprints of raindrops on what was once soft mud but is now solid rock, suncracks have been similarly preserved, as have the casts of worm burrows and the tracks left by other crawling but unknown creatures.

Through millions of years these conditions persisted, swamp deepening to lake, lake shallowing to swamp or even reverting temporarily to dry land, while all the time the land sank slowly and a great thickness of rocks accumulated, shales, sandstones and more shales. The site of Sydney was near the centre of the lake system, and it reached northward to beyond Newcastle, southward to the Illawarra and westward over what is now the Blue Mountains.

This concludes the first stage in the story of Port Jackson. It was then the foundations were laid, particularly the massive sandstones which give shape to the local scenery and on which and of which the city of Sydney is largely built.

There is now a long gap in the story, and some 150 million years must be passed over. Elsewhere great cataclysms took place, new lands appeared above the sea and disappeared again, mountains were thrust upwards and then destroyed, but in Sydney the landscape was unaltered. Both the building of new rocks and the destruction of the old were halted, and the land remained low and featureless, awaiting rejuvenation in a later age. Nothing is found locally to tell the story of this gap, of the countless races of plants and animals that lived for a while, and then died out to be replaced by others. This part of the story must be pieced together from evidence elsewhere, and does not concern our present problem, the story of Port Jackson and its environs.

That brings us to a late stage in the world's history, the events of the last two million years, beginning with the Great Ice Age, during which there is at last evidence of the presence of man on the earth. There has been previous reference to older and even greater ice ages; but this is called the Great Ice Age because it is so close to us, the evidence of its passing is tangible, and the development and migration of early man are closely associated with it. The Great Ice Age practically coincides with the Pleistocene Epoch, and there were several warm interludes during the general refrigeration of the world's climate. The main regions affected by actual glaciation were northern Europe and northern America; its effect on Australia, with the exception of Tasmania, was largely indirect.

During the Tertiary Period Australia, though flat, had a warm and moist climate, great rivers ran through the interior, and the country was fertile and teemed with life. In the east the land was slowly rising. The rising of the land was accompanied by the volcanic activity spoken of in the last chapter. There were volcanoes near Sydney, but not on a grand scale. The worn necks of some of these volcanoes still exist, filled with solidified lava and broken material from explosions. One of the outlets was near Dundas on the Clyde-Carlingford line, and the quarry from which road metal was formerly obtained is within the old vent. Another vent was near Hornsby. The remains of sheets of lava from other unknown craters now cover Mt Wilson and other high points on the Blue Mountains.

It was in the first stage of the rising of the land that the valleys of Port Jackson, Broken Bay and Port Hacking were gradually excavated. The plateau at this time was about from 700 to 1,000 feet above sea level, the highest part in the north. It is doubtful if any considerable river ever flowed into Port Jackson; certainly not after about the middle of the Tertiary Period, when the land commenced to rise. The main drainage was to the west and into the river which still flows as the Hawkesbury into Broken Bay.

High land always means the beginning of destruction, the breaking down of the rocks into sand or mud and their removal by running water. The valley in which Port Jackson now lies was so cut through the underlying sandstone to a depth of about 600 feet. The valley of the Hawkesbury to the north was deeper, about 900 feet. Having reached a depth of 600 feet, the sluggish stream at the bottom had little fall and could no longer carry away the material worn from the hills. Its excavation of the valley then ceased.

It is harder to picture the life of mid-Tertiary times than the topography. Examination of pollen in ancient deposits of the River Lachlan in the western slopes of the Eastern Highlands has shown a high proportion of rain-forest pollen—of beeches like those of the New Guinea highlands, of *Podocarpus* and of tree ferns. But there are also pollen of the more typically Australian bush, of Myrtaceae, Proteaceae and Casuarinaceae. These increase relatively higher up in the sediments and so are later in time. There may have been waratahs, banksias, Christmas bush, boronia, Christmas bells and flannel flowers; but it is more likely that their representatives were the ancestors of those living at the present day. It is the same with the animals. Wallabies and kangaroos and other familiar marsupials lived in many extinct forms during the Pleistocene, and probably similar or allied forms lived in the Tertiary; but until there is definite fossil evidence this question must remain unanswered. Man alone as yet had no place in the scheme of things. It was probably a good deal later before very primitive man-like creatures made an appearance in central Asia or Africa, and it is certain that a long time would elapse before the first Australians reached these shores.

In the later part of the Tertiary Period the upward movement of the eastern coast near Sydney quickened, but it was more complicated than in its first phase. Instead of rising evenly it remained nearly stationary on the eastern side and sank appreciably in the centre in the vicinity of Penrith. West from Penrith the land rose to form the main tableland of the Blue Mountains, more than 3,000 feet above sea level. This was the genesis of those scenic masterpieces of which more will be said in the next chapter.

The elevation and folding that formed the Blue Mountains may be traced in several ways. As the railway begins to ascend the plateau the sandstones in the cuttings may be seen dipping sharply to the east instead of lying horizontally, showing that they were bent down as the land sank on the eastern side and upwards as it rose to the west. Near here also and on the hills near Wallacia, river gravels, formerly in the beds of running streams, may be seen on steep hillsides in positions where no streams could now possibly run. In places the elevation was so slow that the rivers were able to keep pace and cut their channels as fast as or faster than the land rose. The gorge of the Hawkesbury just south of Penrith was so cut, isolating an outlying spur of the main elevation.

Another way of measuring the relative elevation is by tracing the same rock formation or horizon from place to place. Such horizons are difficult to find in the sandstone, for this formation is remarkably homogeneous throughout its maximum thickness of over 1,000 feet. Either the base or the top of the formation may serve as a guide, for it is assumed that both were originally horizontal though now at different levels. Thus the base of the sandstone is now at sea level at Narrabeen, six miles north of Sydney Heads, it is 890 feet below sea level in the No. 1 Cremorne bore, 994 feet below sea level in the Balmain Colliery, 825 feet below sea level in the Penrith bore, about 2,800 feet above sea level at Mt Victoria, sixty miles west of Sydney, and about 1,500 feet above sea level sixty miles south of Sydney.

This shows that the valley of Port Jackson is at the bottom of a basin, with even rims to the north and south, a high western rim, and a low eastern rim facing the sea. The slight tilting of the eastern rim levelled the floor of the valley, so that instead of sloping to the coast it was probably actually deeper in its western recesses. The over-all subsidence of the eastern part of the tableland now submerged Port Jackson for the first time beneath the sea. This was but the first of several submergences. The section shown in Figure 13 illustrates this and also the sequence of following events.

The story of Port Jackson through the Pleistocene Epoch or Great Ice Age follows a pattern which may be traced throughout the world. The main factor was the eustatic movements of the sea caused by the growth and shrinkage of the great ice sheets which from time to time covered vast areas. Remnants of the ice sheets still remain in Greenland and Antarctica, many thousands of feet thick and covering several millions of square miles.

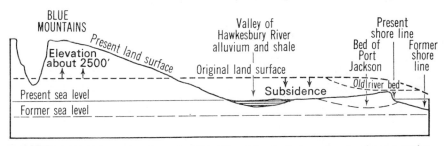

13.   Diagrammatic section from the Blue Mountains to Sydney showing how elevation in the west and subsidence in the east causes a central basin and how the final change in sea level produced the present harbour.

It is only in recent years that the effect of this enormous mass of ice upon the level of the oceans has come to be understood. Just prior to the Great Ice Age world climate was generally warmer than it is now; although there were ice sheets in Greenland and Antarctica, they were small. If all the ice at present in Greenland and Antarctica were melted and added to the ocean it would be sufficient to raise sea level no less than 130 feet. Evidence that sea level was once nearly as high as that is found in many places, in Australia as elsewhere, in the form of raised beaches. Raised beaches are terraces of sand and shingle well above sea level, often containing shells and other remains similar to those in the neighbouring seas. They invariably show the position of a former coastline.

At the climaxes of the Great Ice Age vast ice sheets covered much of northern Europe and America, reaching as far south as Great Britain and Germany and well into the United States. There were warm interludes between periods of intense glaciation, and the ice sheets shrank and grew again many times. All the water forming the ice came directly from evaporation over the sea, and its alternate melting and re-formation caused many fluctuations in sea level.

There were at least four glaciations, probably more, when sea level fell well below its present level because of withdrawal of water from the oceans to form the ice sheets. But it has proved difficult to work out with certainty the course of sea-level change during the earlier glaciations and the intervening warm interludes, and geologists disagree strongly about this. The difficulty arises from the fact that the land itself has moved up and down as well and the farther back in time one goes the harder it is to disentangle movements of sea level itself from movements of the land. However it is reasonably clear what happened from the last warm phase in the Pleistocene prior to the final formation of large ice sheets. Many of the raised beaches at about 100 feet belong to that warm interval so sea level was then at that height above its present position. Then during the final great accumulation of ice, the sea went down at least 400 feet below its level today, maybe as much as 475 feet.

During all this time Port Jackson suffered many vicissitudes. At one stage it was an even more extensive harbour than it is now; at another, when the sea receded, it lay as a peaceful valley with a sluggish stream meandering through shallow swamps in which beds of peat formed.

This brings us close to the present day. The last great melting of the ice, and with it the return of water to the oceans, began well before 10,000 years ago, the date usually taken for the start of the Recent. This return of water was virtually completed by about 6,000 years ago, by which time the coast was very like it is now. Many geologists think we are even now within the Ice Age, passing through a warm interlude of uncertain length. The future may again see the formation of ice sheets over large areas of temperate land and another lowering of sea level throughout the world. On the other hand there is evidence that for the time being world climate is getting warmer, and that there has been a definite shrinkage in the ice sheets of Greenland and Antarctica and in the glaciers of Iceland, Switzerland and elsewhere, all within the last 50 or 60 years. Another thing which seems certain is that about 3,000 years ago there was a temporary chilling which added to the ice sheets of Greenland and Antarctica. This is known as the Little Ice Age. Whether this enlargement of the glaciers was great enough to have an appreciable effect on sea level is uncertain. Some investigators find evidence on certain coasts of emerged shorelines about 10 feet higher than today's and attribute them to this last small expansion of ice on the land. But others find no such evidence in other areas, sea level in this period being the same now or even rising slightly to the present level. It may be there was no eustatic sea level higher than the present in the Recent, and local crustal warpings are responsible for the evidence which has been taken to point to such a Recent high sea level.

Tracing the changes in topography of Port Jackson during the Recent Epoch brings to light many other points of interest. They are here dealt with in some detail, for the same principles, though varying slightly, may be applied to practically all harbours and bays, not only upon our own coast, but elsewhere in the world.

Looking at the map (Figure 14), we see both North Head and South Head as peninsulas, each connected to the mainland by a low-lying isthmus. Both peninsulas are high and rocky. Just within the entrance to the main harbour is a shallow bank called the Sow and Pigs Reef. The bank consists mainly of sandy mud, but in several places there are rocks which are just awash at low tide.

The particular point about the Sow and Pigs Reef is that its position in this locality leads to the interesting hypothesis that instead of one original river valley there were really two, each with different outlets to the sea. One of the valleys is now Middle Harbour, and its outlet was between the present heads. The other valley, now the main harbour, was separated by a low col,

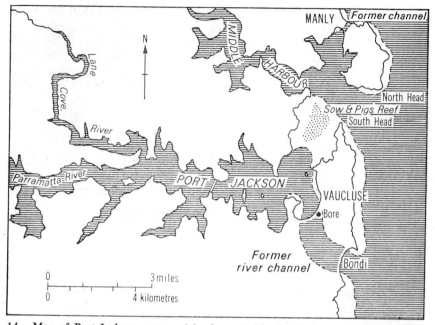

14.   Map of Port Jackson, now and in the past. The hatched area shows its extent when North Head was an island and there were two harbours with separate entrances to the sea.

now the Sow and Pigs Reef, which joined higher land from George's Heights in the west to Vaucluse Heights in the east. The outlet of the stream flowing through this valley was through the gap between Bellevue Hill in the south and Vaucluse Heights in the north. This gap is now low-lying and connecting Rose Bay to Bondi Beach, and there are golf links and a residential area where previously there were sand dunes and small swamps.

In seeking evidence of the existence of two separate valleys a great deal depends on the comparative depths of bedrock at Sow and Pigs Reef and between Rose Bay and Bondi. Sow and Pigs Reef lies athwart the shipping channel, and in the early days of settlement the maximum depth of water above it was only about 20 feet. This was quite sufficient for the sailing ships of the times, but the needs of modern shipping are much greater. Both the eastern and western channels have now been deepened by dredging to 45 feet. Actual dredging scooped out holes rather deeper than this without reaching rock, but there is evidence that bedrock was not far below the lowest point reached.

The dredging of the eastern channel some years ago by the suction dredge *Triton* provided interesting information, as well as a harvest for local shell collectors. The material raised by the great suction pumps at the rate of over 2,000 tons an hour was not as usual dumped at sea, but was taken by punt well up the harbour and used for reclaiming swamp lands at Dundas. Thus spread out it afforded a wonderful picture of the wealth of shell life in our waters. Some of the finds came from a solid sandy bed some depth below the surface; they included many tropical sea shells unlike those living in the harbour. Some are apparently extinct, but others are related to species living on the coast of north Queensland and in New Caledonia. These specimens are apparently of great age, but they have retained their original lustre, though the colour has faded uniformly to a bright yellow or orange.

More significant were specimens of coral of a reef-building type, all found in the lower levels. These must have grown on some solid foundation such as rock, and there is of course the possibility that they had been washed from the rocky sides of the channel into the positions where they were found. On the other hand their condition suggests that they were very nearly in the place where they had grown, that is, perhaps 70 feet but not more than 80 feet below present sea level.

When the western channel was dredged for the passage of the Queen Mary during World War II, the dredge reached to about 50 feet below sea level. Here the sediments were very coarse and there was much broken rock. Unless this had been a river channel in a very narrow gorge, the broken rock suggests that (a) bedrock again was not far down, and (b) there had been originally a continuous ridge of rock connecting the high land on either side of the harbour, and at a maximum depth of between 60 and 80 feet below present sea level.

However recently, geophysical methods have been used to sound the sediments between George's Head and Vaucluse. Electromagnetic waves are used to penetrate the materials beneath the harbour and be reflected back; their time of travel enables the depth of the bedrock to be determined. This work has shown a considerable thickness of loose sediment here which has accumulated since the valley was flooded by the sea.

In Rose Bay the ultimate depth of bedrock is not yet known. A few bores had been sunk in search of water, the deepest until recently on the Rose Bay Golf Links. This reached 91 feet without striking bottom. Then another bore was sunk, and through the courtesy of Mr Royle the contractor I was able personally to watch its progress. The site was at Lyne Park, Rose Bay. The bore passed through various layers of marine sand and mud, peat and shale. By washing mud from depths of 75 to 95 feet, many fragments and small shells were obtained, mainly identifiable with those still living in the harbour. The bore bottomed on sandstone at 148 feet, or 143 feet below mean tide mark. As the bore is not in the centre of the low ground, it is probable that it is to the side of the old valley, the bottom of which would lie a little to the south.

Farther up the harbour and above the bridge the depth of bedrock is about 150 feet, so that even without allowing for the backward tilting of the valley during folding, the drainage could have followed the slope of the ground and passed to the sea beneath Rose Bay and Bondi Beach. However the evidence is not conclusive and the hypothesis of two separate valleys cannot be regarded as proven.

During the various inundations of the sea the harbour gradually filled with mud and sand. Many think that Port Jackson is a deep body of water; actually it is comparatively shallow, with a flattened floor, averaging about 50 feet below mean tide level. There are a few deep holes, the deepest one above the bridge reaching practically to rock bottom at 150 feet. This means that since the sea first entered the harbour, some 90 feet of sand, mud, shells and other marine organisms have been deposited. In the bore at Rose Bay beds of peat between sand beds containing sea shells mark the stages when the harbour became temporarily dry land again.

Let us try to picture the final inundation by the sea. Whether there were two valleys or not, as the sea rose, it would eventually flood through both the Rose Bay-Bondi gap and the Sow and Pigs Reef ridge to make Vaucluse Heights an island and join Middle Harbour to Port Jackson. Similarly the sea passed through Manly and made North Head an island.

Off the coast there is a strong, persistent ocean current called the Notonectian Current flowing from the tropics down the New South Wales coast to the south beyond Sydney. It carried larval forms of many tropical marine animals, molluscs and corals among others, during the submergence and many of these came to rest in the shallow water at Sow and Pigs Reef. Conditions there were favourable for coral growth, with clear warm water, continually renewed by the tropical current. It was then that the corals in the east channel flourished, as did the tropical shells found in the *Triton* dredgings. Even today there are many tropical marine creatures brought by this current which find a precarious hold upon our shores. Within the harbour tropical shells are found at Bottle and Glass Rocks, and at Watson's Bay there is still living beneath low tide a large mass of coral about seven or eight feet across. The southern limit of the current is about Shellharbour, 60 miles south of Sydney.

Tides carried mud and sand into the southern gap at Bondi. Then waves built sand into a natural barrier across its mouth. On top of this wave-constructed feature the wind formed dunes which existed at Bondi until a few years ago.

The gap at Manly filled similarly but remained open until much later. Even in the early days of settlement a narrow channel in the vicinity of what is now Ashburner Street connected the harbour with the ocean. The local fishermen have an interesting theory about this channel. At certain times of the year schools of kingfish migrate north along the coast, and entering the harbour congregate in the corner of Manly Cove where the entrance to the channel formerly existed. Here they are netted in great numbers. The theory advanced is that this was their original passage to the sea, and that inherited instinct still urges them to seek egress through a channel that no longer exists.

With the closing of the gap at Rose Bay conditions deteriorated for the growth of coral. The reduction in volume of open seawater flowing across the reef would restrict growth. The deposition of mud and sand finally killed the coral altogether.

The filling of the harbour with mud and sand was very slow. The upper portion of the 90 feet of sediments has been deposited during the last inundation of the sea, the lowermost probably before the commencement of the Great Ice Age. It is difficult to estimate the rate of deposition, or to say what proportions belong to the different inundations. An estimate was made by the Maritime Services Board of one inch a year, based on observed results in dredged channels. But this was seen to be local and temporary, and due largely to the redistribution of existing sediments by tidal currents. Actually it must be much slower.

In the period of the white man's occupancy there has been no discernible shallowing of the water. This conclusion has been reached in the only way available, by comparison of the early charts with those of the present day. This is not quite satisfactory as it is impossible to pinpoint any locality, and there is no proof that both are measured from the same datum line. With this in mind the following comparative table should not be taken as conclusive. The figures given are for the greatest depths within limited areas, and are taken from Captain Freycinet's Atlas of 1812 and from the latest charts available, thus covering a period of a little over 150 years.

TABLE OF COMPARATIVE DEPTHS

SITE	DEPTH IN FEET	
	1812	TODAY
Middle Harbour, above the Spit	96	Over 100
North Harbour, due west of Quarantine Point, centre of channel	54	50-60
North Harbour, to side of channel	42-48	40-50
Main Harbour, off Bradley's Head	60	Over 60
Main Harbour, north of Pinchgut	60	Over 60
Main Harbour, W-N-W of Bottle and Glass Rocks	102	Over 100
Main Harbour, above Bridge, greatest depth	138	150

It would seem from these figures that the harbour in 150 years has actually deepened rather than silted up, but it is not asserted that the figures are accurate enough to warrant this conclusion. They do show at least that sediment is being deposited so slowly that in our own time there has been no appreciable difference in depth.

A few miles to the north of Sydney the road follows a narrow, coastal plain with high hills rising to the left. At Narrabeen there is a series of large lagoons stretching backwards for some miles to the foot of the hills, and intermittently connected by a narrow channel with the sea. Beyond Narrabeen the plain winds through Warriewood, with hills also on the seaward side, to end at Pittwater, a long arm of the sea stretching southwards from Broken Bay. The chain of hills separating Pittwater from the main ocean is broken by several gaps, all but little above the present sea level.

The whole of the plain, the lowlying land between the hills and the floors of the lagoons, was originally below sea level, and represents a platform excavated by the breakers. The hills were thus islands which are now joined to the mainland. Sea shells may be found in any excavation on the low land, shells of exactly the same species as those now living in the neighbouring sea. The old coastline is the foot of the main line of hills to the west. Then the great waves from the south-east sweeping across the Tasman Sea washed sand up to form beaches and barriers linking the headlands and the islands. In this way the lagoons were formed.

It is a peaceful corner of the coast, sheltered from violent winds and almost tropical in its verdant richness. The sandy soil has been enriched by vegetable matter washed from the hills and from the decomposition of the Narrabeen shales, which contain much volcanic ash. Palms grow in abundance, and bananas and other tropical fruits flourish in the rich soil.

The story of Port Jackson is the story of many other parts of the coast of Australia. It is told here at some length because Port Jackson was the first settled part of Australia, and because the presence of a great city so often blinds the eyes to the significance of the greater story beneath its foundations.

The general rising of the sea after the last great melting of the ice produced many other harbours besides Port Jackson. Some are drowned river valleys, others such as Botany Bay, Jervis Bay and Port Phillip are low-lying areas submerged when the sea rose. Tasmania itself became separated from the mainland as did New Guinea in the north. The coast of Queensland was greatly affected, but here there may have been later subsidence as well. This is closely interwoven with the story of the Great Barrier Reef, the subject of another chapter.

The Blue Mountains

To the west, from some of the higher points around Sydney, the view is over and beyond a region of low undulating hills. Beyond this again on clear days there can be seen just above the horizon a narrow even ribbon of blue. In a few places minute rounded protuberances break the otherwise level line. In the late afternoon the shadow of the sinking sun turns the blue ribbon to black; then it fades into the growing dusk and is lost. That blue ribbon is the plateau of the Blue Mountains, some 30 miles distant across the coastal plain, and the protuberances are Mt Wilson and a few other basalt peaks that rise a few hundred feet above the general level.

As one goes westward by road or rail the mountains are rarely glimpsed until the hills forming the divide at the head of the Parramatta River are passed and the flood plain of the Nepean River is reached. The river here flows from the south and skirts the base of the mountains until it turns east into the long winding estuary of the Hawkesbury River and Broken Bay. At closer range the mountains lose their blueness and appear as forest-clad slopes terminating abruptly at the edge of the alluvial flats of the Nepean.

Though not of great height, this barrier for many years halted the westward expansion of the young colony. The story of the first explorers who tried to find a way through the mountains has often been told, and as this is not a narration of human events, it is sufficient to summarize the many attempts. It was natural that these at first should be by way of the valleys. Experience in other lands showed that mountain passes were usually at the lowest points, often a col between high peaks at the head of a valley.

Those who sought to penetrate the Blue Mountains through any of the numerous gorges were soon disillusioned. The gorges invariably remained narrow and steep, flanked by beetling cliffs of solid rock, their bottoms choked with huge masses of sandstone fallen from above. Over all was a tangled mass of vegetation, trees clinging precariously to precipitous slopes, vines and undergrowth hiding dangerous pitfalls. The bottoms of the valleys yielded no practical path, and on the steep slopes at the sides ledges faded out into the face of vertical cliffs, forcing a retreat just when progress seemed possible. A few miles took days and even weeks to negotiate, and with every step the difficulties increased. Rarely has such a low barrier, its highest point little more than 3,000 feet, so defied the efforts of the explorer.

The way when found was comparatively simple. It lay over and not through the mountain barrier. When Blaxland, Lawson and Wentworth determined to follow the ridges they pioneered the passage to the interior and ultimately to much of the future wealth of Australia. The pass they found is still the main route of road and railway, a route from which there can be little deviation. After ascending the first slopes near Glenbrook, they followed a long ridge upwards to the plateau top at Wentworth Falls. Sometimes the ridge widened to a mile or more, sometimes it was little more than a knife edge, with precipitous slopes leading down to the depths on either hand. On this ridge trains now roar through deep cuttings or over high embankments, and close to it the road twists and turns, first on one side, then on the other, and villages have sprung up wherever there is room to build. Apart from this much of the country is as it was before the advent of man, and in the gullies the bush is still virgin and untouched, save where devastating bushfires, the result of man's carelessness, have left their scars.

The upper level of the tableland is reached at Wentworth Falls. For a few miles farther on the level rises until it is over 3,000 feet above sea level, but the rise is slight and not very noticeable. The first glimpse of the summit is misleading. There is nothing to contradict the expectation of a wide and elevated plain. The illusion is continued as the road leading from Wentworth Falls railway station is followed for about a mile to the south. The forest has been cleared, the land is level, streets branch to right and left, and lines of cottages suggest suburbia rather than primitive wilds.

Then suddenly the road ends, and the world is cut off at one's feet. The abruptness of the change is breathtaking. We are on the brink of a vast chasm, the misty depths going down and down until mighty trees far below appear as mere shrubs. The walls of sandstone are sheer for over 600 feet, and at their base steep slopes of talus continue to the stream at the bottom, its rushing water hidden from here by the overhanging foliage. It is not more than two miles as the crow flies to this stream, but it is over 2,000 feet below. On either hand the walls of sandstone continue, sometimes jutting out, sometimes receding, like the capes and bays of a rock-bound coast. Directly opposite is another line of cliffs crowning the great mass of Mt Solitary. To the left the valley opens downwards to a distant vista of forest and valley to where the Cox River is hidden in a narrow cleft less than 300 feet above the level of the sea.

There are times when the bottom of the valley is not visible. Often in the the early morning mist and cloud fill the vast chasm, and the eye looks out over the surface of a white sea. The cliffs opposite and on either hand, rising everywhere above the level of the clouds, complete the illusion of a lonely sea shore.

This is the Jamieson Valley, one only of the great gorges forever eating into the heart of the tableland. The tableland, which from afar seemed so solid and substantial, is even now but a shell. From the air or studied on a contour map, the valleys are seen to cover by far the greater part of the area, and the high land is confined to a pattern of narrow ridges separating the basins of the various streams. The ridges are but the remnants left by the relentless chisel of nature, and even these are in process of dissolution.

A few solid pieces of the tableland still stand. To the left of the Jamieson Valley, and forming its eastern boundary, is King's Tableland, a long sandstone-capped spur running southward to its terminating point above the gorge of the Warragamba River. Across the Jamieson Valley to the south is Mt Solitary, a great isolated mass of sandstone about five miles long, once part of the main tableland, but now only connected on its western side by a low ridge.

On the western side of the Jamieson Valley is another wall of sandstone, seemingly the edge of a wide tableland beyond. But this again is a fragment, a mere ridge extending outwards and separating Jamieson Valley from Kanimbla Valley to the west. A path from Katoomba leads to this ridge across the col at Narrow Neck, and here the vertical sandstone cliffs separating the two valleys are only about 100 yards apart.

The largest fragment of the tableland is the main ridge extending north from Wentworth Falls. On this ridge cluster the villages and towns of Leura, Katoomba, Blackheath and Mt Victoria. It is here and there a little wider, but again there is little space on either side, for while the Jamieson and Kanimbla Valleys flank it to the west, the equally imposing gorge of the Grose River encroaches upon it from the east.

Blaxland, Lawson and Wentworth came this way. Once committed to the ridge there was indeed no other. Westward the sandstones begin to thin out, and the cliffs, though still imposing, are not so high. Here and there a way may be found to the valleys below, and at Blackheath a road has been cut down the side of a long ridge into Megalong Valley. Tracks have also been carved down the cliffs in a number of places, giving visitors access to the waterfalls and to the rain forest of the slopes and valley floors. The men who made these tracks were men of iron nerve. Lowered by ropes they made their way from ridge to ridge, digging paths while balanced precariously over giddy depths, often cutting steps in the face of the cliff itself.

The first explorers went a good deal farther before they were able to descend. The route they found is still the most practical, and is followed by all road traffic across the mountains. This is the Victoria Pass near Mt Victoria, and the road follows a spur downwards to the Lett River, itself a tributary of the Cox. Beyond Mt Victoria the sandstone ramparts begin to break down, the last of them being Hassan's Walls near Lithgow. The sandstone escarpment is here much broken, but Hassan's Walls still tower majestically above the valley of the Cox, and overlook to the west the broken country leading to the main divide.

From many points along the main ridge, where the view is clear to the north-east, a few rounded heights may be seen across the Grose Valley, heights rising a few hundred feet above the general level of the tableland. These are the heights that had appeared from the eastern distance as mere knolls breaking the even horizon of the mountains. Mt. Irvine appears right across the valley, Mt Wilson lies a little farther to the west. Geologically they have their own special interest.

Just when and where volcanoes were active in this part of the country is not known. It was certainly before the excavation of the gorges, and may well have been before or during the general elevation of the tableland. As already described, volcanic activity had occurred somewhere or other in eastern Australia right through Tertiary and up to Recent times, and this no doubt was part of that general unrest. Somewhere in the vicinity volcanic vents opened up and floods of lava found an outlet, covering quite a large area to a depth of two or three hundred feet. The volcanoes have long since become extinct, even their position is unknown, and of the lava-flows only a few scattered remnants are left. It is these remnants that form the few higher isolated peaks.

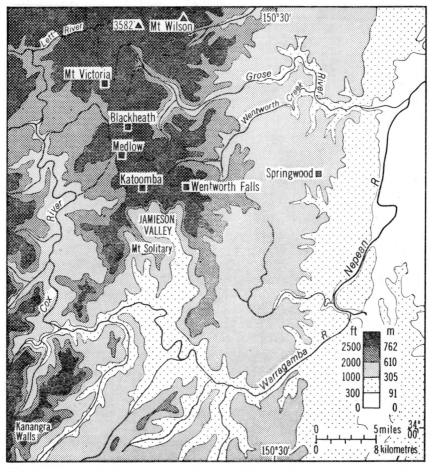

15. Map of the Blue Mountains showing how little remains of the original tableland.

They are veritable oases on the rather arid summit of the tableland. The rich soil formed from the decomposition of the basaltic lava has allowed a profuse growth of vegetation of the jungle type. There are not only such trees as sassafras and coachwood, but the fauna is also typical of the rain forests of the tropical north. In the warm moist climate of pre-glacial days the rain forest spread far beyond its present range and even well into the interior. In clays buried beneath the lava-flows here and elsewhere impressions of leaves and fruits may be found quite unlike those now living in the vicinity, but similar to those of the jungles of Queensland and New Guinea. A few remnants of these jungles still survive in outlying parts, and those on the Blue Mountains are very nearly at their southern limit.

In striking contrast to the jungle is a little plant clinging to the rock ledges right at the bottom of Wentworth and Leura Falls. This is a small drooping shrub that grows only where it can catch a little spray from the falling water. It has already been mentioned in the chapter on the Australian bush as a member of the pine family called *Microstrobos,* and it is now very rare. This is a different species of *Microstrobos* from that found on lonely mountain tops in Tasmania.

The river system of this part of the country is complicated and at first sight confusing. The first western descent from the sandstone plateau, by road at the Victoria Pass and by rail at Lithgow, is still within waters that eventually find their way to the east. The main tableland continues well to the west, but variation in hardness of many geological formations has led to a confused mass of hills and valleys, and it lacks the regular structure of the sandstone-capped portion to the east.

Both road and railway, after leaving Lithgow, wind through a narrow valley draining into the Cox River, then over the headwaters of the Cox itself at Wallerawang. The Cox River is here a small narrow stream nearly 3,000 feet above sea level, and just about to plunge into a deep gorge beneath the shadow of Mt Walker. Mt Walker is a cone of quartzite 3,896 feet high, a conspicuous landmark from all sides; it shows how the original elevation of the main tableland was almost dome-like and here, at about its maximum height, probably well over 4,000 feet.

Between Wallerawang and Rydal the railway line crosses a low range of hills, inconspicuous after the spectacular scenery already passed. Large areas of these hills are now covered with plantations of pines growing for commercial purposes. This is the Main Divide of eastern Australia. Beyond it the tableland begins to slope gently to the west. All the water here flows into streams which feed the Macquarie River, and that which does not evaporate or sink into the ground or become absorbed in the distant Macquarie Marshes reaching the Darling River and thence the distant Murray, eventually to enter the sea in South Australia more than a thousand miles from its starting-point.

The Cox River actually breaks through the sandstone tableland. Flowing in a secondary gorge at the bottom of Kanimbla Valley, it swings eventually to the east through some of the roughest country in Australia. Joined on

Recent rock fall near Narrow Neck, Katoomba, Blue Mountains, New South Wales. *New South Wales Department of Tourism.*

Hinchinbrook Channel and Mount Diamantina, Hinchinbrook Island, Queensland. *Queensland Department of Tourist Services.*

Mulgrave River "corridor" between the Bellenden Kerr Range on the left and the Thompson and Graham Ranges on the right. *F. W. Whitehouse.*

Lake George, New South Wales, with the Cullarin Range scarp on the left.

the south by the Kowmung River, which rises near Kanangra Walls, and on the north by waters draining Jamieson Valley, it then unites with the Warragamba River. The Warragamba, flowing through a narrow gorge near the edge of the tableland, is in turn joined by the Wollondilly, which flows through Burragorang Valley from the Southern Tableland near Goulburn. This valley is now flooded by the Warragamba Dam built in its narrow mouth prior to its emergence onto the Cumberland Plain. The combined streams then unite with the Nepean near Wallacia, and they bear this name until tidal waters are reached. The long estuary, a drowned river valley like Port Jackson, ends in Broken Bay and is called the Hawkesbury River.

It is the sandstone capping that gives character to the Blue Mountains and defines their geographical limits. The Blue Mountains end at Lithgow, but the main plateau extends another 100 miles to the west, until beyond Orange it begins to descend steeply to the great plains of the interior.

The story of the formation of the Blue Mountains is largely that of Port Jackson as told in the previous chapter. It may be divided into three phases, the actual formation of the rocks composing the tableland, their elevation into a tableland, and the dissection of the tableland into the tremendous gorges that now exist. In the chapter on Port Jackson the Hawkesbury sandstone was taken as the foundation on which the present scenery is built; in the mountains it is necessary to go below this to formations of a much greater age.

If any of the tracks from the summit be descended, beneath the masses of soil and débris fallen from above, beds of soft shale will be found here and there outcropping at the base of the sandstone. The shale is of no great thickness, and it is contemporary with the shale beds outcropping on the coast north of Sydney at Narrabeen and elsewhere. Beneath the shales are the coal measures—coal seams, shale and sandstone deposited in the swamps and lakes of the Permian Period. These coal seams are worked at Lithgow just below the western edge of the Blue Mountains; they are at sea level at Newcastle; they are tapped in the Balmain Colliery nearly 3,000 feet below the surface at Sydney; and they rise again to outcrop on the face of the Illawarra Mountains on the south coast. Below the coal measures again are the thin edges of a large sandstone formation containing fossil sea shells, and though these are older they still belong to the Permian Period. Underneath all and right at the bottom of the gorges is a pavement of hard rocks belonging to still older periods, a pavement formed by the worn bases of what in a remote age were towering mountains. The pavement may consist of granite, exposed in the beds of the Cox River and its tributaries, or it may be of quartzite, a hard siliceous rock that was once sandstone, but was later consolidated under great heat and pressure. The quartzite often contains the casts of sea shells which lived in the Devonian sea more than 360 million years ago. The originally horizontal beds of the quartzite are now tilted on end or contorted into complex folds, evidence of the great forces which once compressed them into mountains long since destroyed.

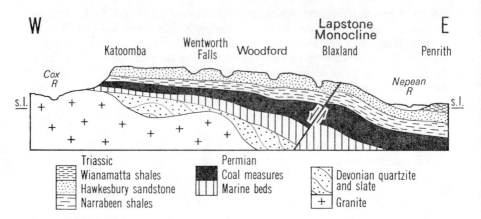

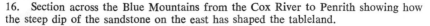

16. Section across the Blue Mountains from the Cox River to Penrith showing how the steep dip of the sandstone on the east has shaped the tableland.

Such rocks form the summit of Mt Lambie on the main divide a few miles to the west. In them and associated with the fossil sea shells may be found fragments of a plant which had evidently drifted from the neighbouring land to the position where it was finally buried. This plant is of great interest, for it formed what were probably the first forests on the face of the earth. Needless to say it was of very primitive form, without flowers, and even a little lower in the botanical scale than the ferns that were its contemporaries. The main character by which it can be recognized is the pattern of the rhomboidal scars left on the trunk and branches as the leaves became detached. It is mentioned here chiefly because, though practically extinct for long ages, one of its descendants still lives upon the Blue Mountains. In the heath country near Katoomba it may be found in clusters, a small erect plant known as "club moss", not more than a foot or so high, but similar in most of its characters to the extinct *Lepidodendron,* which once predominated in the plant life of the world.

Such in brief are the strata composing the Blue Mountains, and the long and intricate story of their formation is the first phase in the formation of the present scenery. In the previous chapter it was told how for long ages the land had lain undisturbed at a low level. It was in the Tertiary Period, that the second phase began. This was the elevation of the mountains, an elevation affecting the whole of eastern Australia to some extent.

We have already seen how this elevation at first raised the land about 1,000 feet and extended right to the coast in the vicinity of Sydney. After the first elevation the excavation of the gorges began, and the valleys were cut which were later to contain the waters of Broken Bay, Port Jackson and Port Hacking. The valleys at this stage were comparatively small, a few hundred feet deep and making little impression on the bulk of the tableland.

Then came the major elevation, a tremendous tilting of the surface with a depressed fold to the east, so that many of the eastern valleys sank below sea level and became harbours. The elevation lifted the Blue Mountains over 3,000 feet above sea level, and a little to the west the top of the dome was over 4,000 feet. This was the last great movement of the land in this region, and it left the height of the mountains substantially as it is now. It must not be thought that this great movement was necessarily sudden or cataclysmic. Judging by the way the Nepean River on the margin was able to cut its bed downwards in time with the uplift the rise was very slow and occupied a long period.

The last and most spectacular phase now began, and is proceeding at the present time. This was the dissection of the tableland and the excavation of the gorges to their present depth. It is hard to realize that the principal agent in this vast destruction is that most simple one, running water.

All philosophical conception needs the power of imagination. Without imagination the concepts of time, space and matter are limited within the range of personal experience and perception. If an insect could think, its span of life of a few days would seem normal, and by comparison a man's life of 70 years would seem infinitely long. A microscopic organism, given the same power of thought, would find the bulk of a whale far beyond its conception. Man, with his more highly developed mind and with growing capacity for mental imagery, is to some extent fitted to grasp the magnitude of events beyond his own experience, but even he soon finds a limit to conception, a limit narrow indeed compared with the infinity beyond.

Knowledge of natural laws and processes has widened greatly in recent years, which may explain why even so great a scientist as Charles Darwin could not visualize the excavation of the gorges by ordinary agents of erosion. When he stood on the brink of the Kanimbla Valley in 1836 he was amazed by the vastness of the scene and later wrote in his journal: "The first impression is that they have been hollowed out, like other valleys, by the action of water, but when one reflects on the enormous amount of stone, which on this view must have been removed through mere gorges or chasms, one is led to ask whether these spaces may not have subsided. . . . To attribute these hollows to the present alluvial action would be preposterous."

The only solution Darwin could suggest was that the valleys were originally bays in a rock-bound coast, and that they were deepened by the receding sea as the land rose. He thus actually reversed what is now known to be the truth, that is, that the present coastal harbours are drowned river valleys, while the valleys are certainly not the raised beds of former harbours.

Lesser minds than Darwin may be pardoned for failing to conceive the tremendous destructive and carrying power of water, a power increasing with every foot in the height of the land. It is worth considering just how streams do excavate and lower their beds. Even now, when the rivers are far below their original level, the water can carry mud and sand to still lower levels, and when in heavy flood, pebbles and even large masses of rock are rolled along. These not only rub against each other and become rounded and smaller, but they also abrade and cut into the river bed. The largest masses of rock are eventually reduced to sand and mud and washed away altogether. When the tableland was at first raised to about 1,000 feet the valleys were excavated to about 600 feet, but when the final lift was increased to more than 3,000 feet, the rate of excavation must have been greatly accelerated. While the valleys were being lowered by erosion they were also being widened. The amount of this widening depends on several factors, but it is chiefly the nature and structure of the rock formations which determine the ultimate topography.

There are many ways in which rocks exposed on the surface are broken down. One of the most potent agents is frost. The erosion of the Blue Mountains went on right throughout glacial times, and there were long intervals when the climate was colder than it is now. Severe frosts followed by sunshine cause contraction and expansion and shatter even the hardest of rocks. In porous rocks such as sandstone water is absorbed as in a sponge, and at night it freezes and expands and bursts the rocks apart. Roots of plants penetrate between the sand grains and as they grow force them apart. Even burrowing animals play a small part. In the sandstones of the Blue Mountains there is another and more subtle agent of destruction. This is the presence of minute quantities of common salt and magnesium salts in some of the layers. The salts are brought to the surface in solution; in sheltered positions the rain cannot wash them away, so they crystallize and force the sand grains apart. Evidence of this may be seen in rock shelters and caves, and also in some city buildings, where beneath cornices and ledges the sandstone is fretting and breaking away.

Of the rocks composing the bulk of the mountain sandstone is the hardest, and once the capping of sandstone was cut through erosion became even more rapid. Much of the underlying rock is shale, a rock that rapidly disintegrates into the mud of which it was originally formed. This is easily washed down the sides of the valleys and carried afar by the streams at the bottom. When the Nepean is in flood, millions and millions of acre feet of water are yellow with the suspended mud, and each flood carries away an enormous weight of solid material, all débris from the destruction of rocks from the higher levels along its course. And this has gone on not once or twice, nor for a year or so, but throughout the million years and more that have elapsed since first the mountains were elevated.

When the valleys were first formed and before the sandstone was cut through they were V-shaped. The sides, though steep, were not vertical, but similar to the valleys on the lower slopes of the mountains at the present day. Once the sandstone was cut through and the soft shale beneath was reached, the tops of the V became bent inwards, and the valleys were capped by vertical cliffs. This was due to the undercutting of the sandstone by the softer shale. The shale wore back comparatively rapidly, producing cavities and causing the sandstone to overhang. Finally it fractured and fell in great masses into the valleys below, leaving a vertical face behind.

These landslides are still happening, as they have happened continuously throughout the whole period. Most of them take place in remote valleys and are unobserved, but such a fall on a large scale was witnessed a few years ago at Katoomba. For many weeks visitors came to see the gradually widening crack as a huge section of cliff tottered to its fall. None witnessed the final catastrophe as it came late in the night, but many heard the devastating roar, and when the morning came there was a new cliff face about 30 yards back from the old, and thousands upon thousands of tons of shattered rock lay spilled upon the talus slopes below. Since then this fallen material has been slowly disintegrating and working down the slopes to still lower levels. Tracks made across it must be renewed at intervals, for after every heavy rain they become obliterated and overwhelmed by the moving mass.

The undercutting of the sandstone explains one of the problems which puzzled many of the early observers. Many of the mountain valleys are much wider in the upper portions, and when they do eventually break through to the low country it is through narrow gorges sometimes narrowed to absolute chasms. The headwaters of the Grose River, for instance, lie in a huge amphitheatre many miles across, while its outlet narrows to little more than a huge vertical crack. This is explained by the fact that the sandstone is not only much thinner to the west, but it also dips down towards the east, so that streams on the eastern margin have had to cut much deeper to reach the soft shale beneath; indeed they often lie entirely within the sandstone. Near their headwaters, where the sandstone is thinner, the soft shale is reached much more quickly, and after this the valleys begin to widen rapidly.

These are the Blue Mountains as we know them now. By far the greater bulk has been removed, and should present conditions continue, the remainder also is destined to destruction. To us this eventuality is remote; geologically it is in the near future. As the sandstone ridges disappear so also will the last vestige of a flat-topped tableland. The hills will become lower and rounded, their sides less precipitous and smoother, the valleys ever wider, their floors level and covered with alluvial flats. The rivers will no longer surge through narrow gorges, but will meander leisurely on their way to the sea. And man! Perhaps our distant descendants will witness these things; perhaps other races will have taken their place. But of this we need not worry, for it will not come in our time, nor in our children's children's.

The Eastern Highlands

Along the whole of the eastern side of Australia from Cape York to the south of Tasmania there is a belt of country some hundreds of miles wide which in past ages has been a zone of continuous change and violent disturbance of the earth's crust. In a distance of over 2,000 miles there is naturally a great difference in local geology, and as a corollary a great diversity in scenery. Much of this scenery is spectacular; all of it is interesting. To tell in full the story of every part would need in itself not one but many volumes. Fortunately it is possible to discern throughout the threads of one general pattern, and to apply the facts revealed in one locality to others far distant.

In the previous chapter the Blue Mountains were singled out for separate description, chiefly because as a dissected plateau they show so clearly the enormous effects of water erosion. It will now be shown that this factor has also played a tremendous part elsewhere in forming the ultimate shape of the land. There are other factors which have played a part, and these will be dealt with in due course as the chapter proceeds.

It must be again emphasized that throughout the ages the surface of the earth has been in a process of continual change. Great cataclysmic movements caused by intense pressure have elevated the beds of oceans into mountains, and have depressed former lands beneath the sea. Gentler movements of vertical elevation and subsidence, often prolonged for long periods, have had the same general effect.

Australia, as we have said, is an ancient land in which violent earth movements ceased early in geological history. In the east they continued much longer than in the west, but they nevertheless ended a long time ago, probably close on 230 million years. In many other parts of the world the mountain building movements have continued, now in one part, now in another. In New Guinea and in the Swiss Alps they ceased after the Tertiary Period; in the Andes they have continued to the present day.

Only in two parts of Australia have the rocks been folded by earth movements since the end of the Permian. Along the eastern seaboard of southern Queensland particularly about Maryborough a very narrow belt was affected in this way in late Cretaceous time, about 70 to 80 million years ago. At the other extremity of the continent in the neighbourhood of North-West Cape, Miocene limestone has been folded into the anticlinal ridges of Cape and Rough Ranges, probably in Pliocene times only several million years ago.

In eastern Australia since the close of the Permian Period earth movements have been vertical generally, sometimes only local, at other times affecting the whole coast. Before this there were so many changes that the geology is very complicated. In each district it has been the task of the geologist to work out the detail of the various formations, and by comparison with other districts to reconstruct the geography of each particular period. This is far too long a story to attempt here, except insofar as it affects the underlying pattern of the present scenery.

The scenery of the Eastern Highlands is built of several layers. Beneath all are the foundations, a pavement of the older rocks, rocks belonging to several great periods of the world's history which, according to their age, have been subjected to one or several intervals of intense folding. They consist of sandstones, shales, limestones, slates, conglomerates, lava-flows and beds of volcanic ash. They have been variously bent and shattered and their once horizontal strata now lie at all angles with the horizon. One feature common to all these old rocks, particularly in Queensland, is the constancy of their strike, the axes of the different folding movements being invariably in a north-westerly or northerly direction. The pressure which caused the folding seems always to have come from the east or north-east, and the edges of the tilted strata run at right angles to this, giving what Professor Whitehouse calls a "grain" to the country, a grain which has affected the trend of the ridges and the direction of the rivers.

In the course of the folding of the older rocks there were many great intrusions, and masses of molten material were forced into them from below, masses which solidified deep within the earth but were later exposed on the surface by long intervals of erosion. These rocks, owing to their superior hardness, now often form the highest land. Granite is a rock of this type, but it is inconsistent in its weathering. Sometimes it is exceedingly resistant; at other times it decomposes rapidly and the detritus is easily eroded and carried away by streams. The highest mountain in Queensland is composed of granite, as is the bulk of the New England tableland in New South Wales and such mountains as Mt Kosciusko, the highest land in Australia, and Mt Buffalo and Wilson's Promontory in Victoria. However there are also granite lowlands such as that around Bega in southern New South Wales and that of Murmungee near Myrtleford in the Victorian Highlands.

These different behaviours are in part due to variations in the granites. More acid granites, rich in quartz, sodium- and potassium-feldspars, and muscovite-mica, are usually more resistant to erosion than more basic ones, richer in calcium-feldspar and biotite-mica. The more basic minerals are more susceptible to weathering. As important is the size of the mineral grains making up the granite; coarse-grained granites allow weathering agents to penetrate more readily to break up the rock and rot it. The frequency of joints through the rock operates in a like way as does that of cracks between, and even within, the mineral grains. Tough quartz grains sometimes reveal under the microscope the presence of hairline cracks within them. All these permit easier entry of water to decompose the rock and the minerals and make for less resistance to erosional agents. These different characteristics sometimes operate together, sometimes in opposition, making it difficult to fathom the pattern of behaviour of granites as landforms.

Another intrusive rock is quartz porphyry, which is similar in composition to granite, but of a different and closer texture, having cooled more rapidly at shallower depths. Notable among the peaks composed of porphyry are the many isolated monoliths which fill the area between Gloucester in New South Wales and the coast, and the spectacular Mt Warning near the Queensland border, which is visible far out to sea, a familiar landmark to all coastal vessels.

Above the pavement of older rocks are large areas of rocks of later formations. These have been elevated by vertical movements to their present positions, but are relatively undisturbed and often nearly horizontal, just as they were laid down. They include many of the main coal measures of Queensland and New South Wales as well as such formations as the Sydney sandstones and the Jurassic sandstones of Victoria. There are in addition the great lava-flows of the Tertiary and post-Tertiary Periods, the basalts of the Atherton Tableland, the Macpherson Range, New England, Bathurst, the Monaro and eastern Victoria. Some of the basalts are hard and resistant to erosion and their remnants cap many of the higher summits.

It is from all the rocks that the present topography has been moulded. Primarily there is a variation in the way they have withstood the elevation and subsidence to which they have been subjected in later ages, for sometimes they have been warped into gentle folds, at other times they have fractured and become displaced along lines of faulting. In the final stages the shape of each hill or valley has depended on the relative hardness of the rocks, whether they are solid and homogeneous or arranged in well-defined beds or strata, whether these strata lie horizontally or are deeply inclined. Then erosion has determined the details, erosion governed by the elevation of the land and the steepness of the rivers and finally by the climate and rainfall.

In the early portion of the Tertiary Period there was practically no high land in eastern Australia. The whole country had been worn by ages of erosion practically to a plain only a little above sea level. A few remnants of the hardest of the old rocks were all that remained above the general level. There was then a slight elevation varying from a few hundred to not more than a thousand feet, but sufficient to allow for some renewal of erosion.

A fairly reliable picture of the country at the conclusion of this stage may be constructed, for remnants of its landscape still survive in places on the summits of the existing plateaux. It was a country of low relief, the hills low and rounded, covered with soil and with few rocky outcrops. The rivers were sluggish and meandered through broad valleys, their flood plains of alluvium often many miles across. Much of this topography has disappeared since erosion was renewed after subsequent great elevation, but fragments may still be seen in many places, in the districts of Yass, Goulburn, Tamworth and Armidale in New South Wales, at Stanthorpe, the Darling Downs and elsewhere on the Queensland tableland.

Weathered granite rocks on the summit plateau of Mount Buffalo, Victoria.

Blue Lake, Snowy Mountains, New South Wales, partly ice-covered in winter. The lake lies in a Pleistocene glacial cirque. *D. James, Australian National University.*

Blue Lake moraines, Snowy Mountains. Ice gouged out a basin in the rock which the lake occupies, and carried out the debris in trains seen on the right. *D. James, Australian National University.*

Orion Lakes, in the glacially eroded central tableland of Tasmania. The Pleistocene ice sheet moved from left to right abrading much of the rock but quarrying out lee sides. *J. N. Jennings, Australian National University.*

During the Tertiary, prolonged warping with some faulting greatly changed this vast extent of subdued relief. For instance in the first half of this period, east-west subsidence in the south gradually depressed the present Bass Strait area beneath the sea and made Tasmania an island for the first time. At the same time the Victorian Highlands began to be raised up as did the highlands farther north. It used to be thought that most of the elevation to present levels took place in a "Kosciusko Uplift" in late Pliocene and early Pleistocene time. But now it is realized that much more of the uplift must have taken place earlier though it did not create really high land in eastern Australia till the late Tertiary.

The uplift may be likened to a long wave which extended from Tasmania to Cape York, a wave with a depressed trough developing here and there along its eastern border. In New Guinea there was lateral as well as vertical pressure, and rocks which had been laid down in an earlier Tertiary sea were folded upwards into a true alpine range. This central spine of New Guinea does not join the Eastern Highlands but is directed away from Australia through the Solomons, New Hebrides and other Pacific islands to the Alps of New Zealand. These more rigorous earth movements were concentrated into the late Tertiary and Pleistocene. The rising of the land along the eastern border of Australia was slow and varied in magnitude; it was also intermittent, and there were pauses when erosion slowed down, to be accelerated once more as the elevation was resumed. The trough to the east reached its deepest in northern Queensland, where the subsidence of a large section of coast is one of the main factors in the growth of the Great Barrier Reef.

Thus was born the main tableland of eastern Australia, a belt of elevated land, varying in height and width from Cape York to the south of Tasmania. At Cape York, where the tableland first emerges from beneath Torres Strait, it is narrow and low. It rises to the vicinity of Cairns, where it is from 2,500 to 3,000 feet, with the great block of the Bellenden Ker Range towering above the general tableland. Here is Queensland's highest mountain, Mt Bartle Frere, its summit 5,287 feet above sea level. Farther south the tableland is again lower, averaging from 1,300 to 1,500 feet west of Townsville, 1,300 feet west of Rockhampton and about the same in central Queensland and the Darling Downs. The lowest section is also the widest, its maximum width being about 300 miles. Curiously enough, off the coast the continental shelf is here also at its widest, and most geographers think there is a definite connection between these two facts. Between Warwick and Dalveen the elevation rises to about 2,000 feet, the tableland narrowing at the same time, and at Stanthorpe it is 3,000 feet high.

In New South Wales much of the New England tableland is above 3,000 feet and part over 4,000 feet; then it becomes lower again, particularly west of the Hunter River, where there is practically a wide gap. In southern New South Wales the Blue Mountains lie on the eastern margin of the tableland, which is over 100 miles wide and from 3,000 to 3,500 feet high, merging southwards into the even uplands of Goulburn and Yass at about 2,000 to

3,000 feet. South of Canberra the plateau rises rapidly, culminating in the so-called Australian Alps, the roof of Australia, with Mt Kosciusko over 7,000 feet high. In Victoria the highlands bend to the west, still very high, many parts such as Mt Bogong being over 6,000 feet. The tableland continues as the backbone of Victoria westward nearly to the South Australian border, its final bastion the rugged Grampian Range, some summits reaching above 3,000 feet. The Tasmanian highlands are dealt with in a separate chapter.

NORTH QUEENSLAND ISLANDS

Few parts tell so much of the story of the tableland as the mountainous islands which lie close to the north Queensland coast. It may seem curious to select islands for such a purpose, but these islands are essentially parts of the mainland. Their topographic features are visible as a whole as they rise directly from the sea. They are in themselves such scenic gems that they well repay special study.

Hinchinbrook Island, which lies to the north of Townsville, has been described as the most beautiful island in the world. It is about 20 miles long and eight miles wide, roughly rectangular in shape and deeply indented at its northern end. Separating it from the mainland is Hinchinbrook Channel, a placid blue pellucid sheet of water from two to four miles wide, so clear that from the decks of passing ships the white sandy bottom is clearly visible many fathoms below. The western shore of the channel is fringed with a wide belt of mangrove flats, and beyond these on the mainland and on the opposite shore of the island itself the mountains tower many thousands of feet into the sky. Viewed from the seaward side the mountains seem much higher than they really are, for they rise sheer from the sea, culminating about the centre of the island in Mt Bowen, 3,750 feet. They are clothed to the top in an infinite variety of green, from the grey green of eucalypt forests to the deep glossy green of dense jungle, with here and there cliffs and rocky outcrops of red granite and schist which compose the mass of the island. From some distance at sea, the slopes appear deceptively smooth, but the vegetation is very dense, concealing rugged terrain.

Hinchinbrook Channel is itself a feature of exceptional interest. Looking at the map, Figure 17, its possible relationship to the valley of the Tully River is very suggestive. Professor Whitehouse in a personal letter tells me that from the air this relationship is even more obvious, and he considers that both the river valley and the channel are parts of the same valley, the lower portion of which was submerged by the rising of the sea in Pleistocene times. The trend of the valley is also significant, running as it does in a north-west and south-east direction.

The valley of the Herbert River, portion of which is shown in the map, runs in the same direction, and throughout Queensland there are many similar valleys, all with the same trend, between the edge of the main tableland and the sea. These are the Queensland "corridors", a term now in

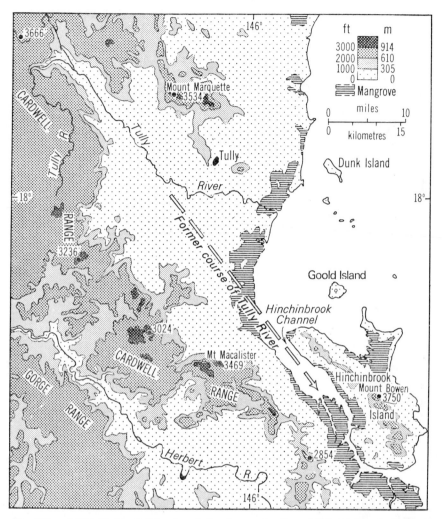

17. Map of Hinchinbrook Island, and the drowned valley of Hinchinbrook Channel and its probable relationship to the Tully River Valley.

general use among geologists. It most effectively describes them, for they have made land communications along the coast comparatively easy, and roads and railways have been able to avoid heavy grades and other constructional hurdles.

The well-known geologist, C. A. Sussmilch, listed no less than ten of these corridors in Queensland, varying in length from 15 to 100 miles and in width from three to 15 miles. All trend in nearly the same direction, that is from 30 to 55 degrees west of north. Some lie directly on the coast, forming a coastal plain; others are deep valleys with the main scarp of the tableland on the west and a high coastal range between them and the sea.[9] The height of the valley floors above sea level varies. Some well inland are very nearly at sea level, others run up to 400 feet, while others such as the Hinchinbrook Channel and the Whitsunday Passage are entirely submerged.

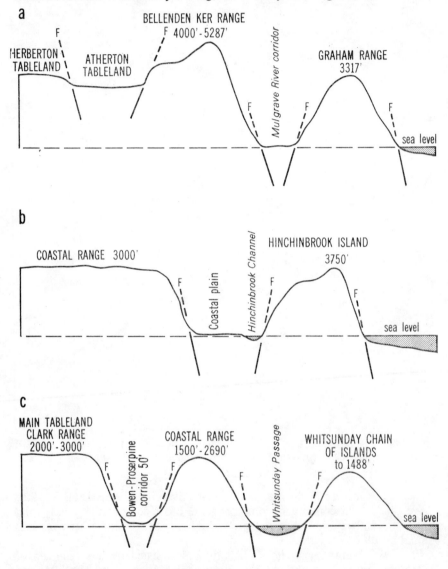

18. Diagrammatic profiles (after Sussmilch) of the edges of the Queensland tableland and some of the "corridors": (a) near Cairns; (b) Hinchinbrook Island; (c) Whitsunday Island.

North of Hinchinbrook Island and just south of Cairns is the corridor of the Mulgrave River in one of the most rugged and picturesque parts of the State. The flat floor of the valley is nearly at sea level, but towering above it on its western side is the great mass of the Bellenden Ker Range, rising almost sheer from 4,000 to over 5,000 feet. On the eastern side of the valley is the equally steep side of the narrow Thompson and Graham ranges separating the corridor from the coast.

The Bellenden Ker Range, though part of the main tableland, is well above its general height. Composed for the main part of granite, it is a great block over 40 miles long and about 10 miles wide. To the south it is also cut off abruptly, and its southern face rises in a steep escarpment directly from the Innisfail plain. On the west face is another escarpment, but this is not so high, rising as it does from the Atherton Tableland some 2,000 feet less in altitude. The Atherton Tableland is partly basalt covered and is itself a region of great interest.

Farther south, near Bowen, where the coast swings away to the east, is very similar topography. Bowen is a picturesque town on the western shore of Edgecumbe Bay, which opens to the north. In the south shallow blue waters cover the broad tidal flats that border an alluvial plain extending far to the south-east in a long narrow corridor a little above sea level. Between this and the coast is a range of granite hills from 1,500 to 2,690 feet high, and on the other side is the main tableland, here called the Clark Range and from 2,000 to 3,000 feet high. The south-eastern end of the coastal range ends abruptly, and here the Proserpine River flows directly across the plain to the sea, fed by tributaries which come in at right angles from either side. On the seaward side is another parallel corridor, the Whitsunday Passage, now submerged by the sea, and beyond this again the partially submerged remnants of another range, now the far-famed Whitsunday Islands, rivalling in beauty and extent Hinchinbrook Island to the north.

The Whitsunday Islands are rather complex in structure, for there are really several festoons of islands, forming portions of distinct submerged ridges. Nearest to the shore is one festoon of islands of which Molle Island and Long Island are the largest, their rocky sides rising directly from the sea, and all densely covered with vegetation. These islands have an interesting geological structure; they are composed of beds of very old volcanic lavas and coarse volcanic ash, which have been folded into a long arch or anticline, its axis running parallel to the shore. The islands form the outer limb of the arch, and the other limb is exposed in the cliffs and slopes of the mainland.

The main chain of islands is beyond the deep channel of Whitsunday Passage, itself another of these curious Queensland corridors. Of the many islands in this group, the best known, from north-west to south-east, are Hayman Island, Hook Island and Whitsunday Island. Hayman Island is not as large as the others but lacks nothing in scenic beauty, particularly on its

[9] C. A. Sussmilch, *Geomorphology of Eastern Queensland,* 1938. Reports of Great Barrier Reef Committee, Vol. 4, Part 3.

seaward side. Here the red granite cliffs plunge sheer down 250 feet into water which is over 20 fathoms deep. Above the cliffs rocky forest-clad slopes lead upwards to the rugged summit 844 feet above sea level.

Hook Island is large, eight miles long and four miles wide. Unlike Hayman Island, it is not composed of granite but of old volcanic rocks, and its jungle-clad slopes terminate in Rocky Hill, 1,310 feet above sea level. On the shore-ward side are flats and mangrove swamps, but on the east the scenery is particularly rugged, deep gullies intersecting the hillsides, while rocky head-lands plunge down in vertical cliffs hundreds of feet to the sea.

Whitsunday Island is like Hook Island but even larger, roughly rectangular in shape, 12 miles long and 10 miles broad. It is exceedingly rugged, com-posed mainly of volcanic rocks, with numerous bays, from the heads of which deep gullies extend far into the interior. Much of the interior is inaccessible. Whitsunday Peak is the highest point, 1,433 feet above sea level. The island is highest in the north, and here its summits are formed of huge rounded masses of bare rock. Bare rock also often forms much of the hillsides, often in unscalable cliffs, though dense tropical vegetation clings wherever it can find a precarious hold.

Farther south near Townsville the Mount Elliott Range is in many ways similar to the Bellenden Ker Range. It is a great wedge-shaped block 25 miles long, 10 miles wide at its south-eastern end and four miles wide at its north-western end, and as a whole greatly exceeding in height the neighbour-ing main plateau of the Leichhardt Range. Mt Elliott, the highest peak, is 4,050 feet above sea level. At both extremities the range ends abruptly against coastal plains. Here again are several parallel corridors at a low level, the Ross River corridor separating the Mount Elliott Range from the main tableland, while on the east is the Townsville corridor, then another low coastal range, then another corridor submerged by the sea, and lastly another ridge running from Magnetic Island to the mainland at Cape Cleveland, where the summit is 1,831 feet high.

A glance at Figure 18 shows the resemblance in cross-section of some of these corridors, and though the sections are diagrammatic they may be taken as representing the structure of many other valleys. Such coastal corridors are not confined to Queensland, and similar valleys are found well down in New South Wales. They tend to show that there is a continuity of structure through the whole length of the eastern tableland.

A splendid example of a coastal corridor is the great valley which runs parallel to the coast from north of the Macleay River, across the Macleay and the Hastings rivers and nearly to the Manning River. The valley is not apparent to the traveller by road or rail, for its margins are ill defined, but throughout its length it is but little above sea level, and it affords a uniformly easy grade for land communications. Here, as in Queensland, there is the main tableland to the west and higher and sometimes rugged land between the valley and the coast.

From the coastal corridors it is a natural step to the main tableland itself, and it may be asked just what happened during the great uplift which gave us our high land. Later in this chapter something will be said of other parts of the tableland, particularly the highest part of all, the Kosciusko Tableland and its western extension into Victoria.

What primarily caused the uplift is debatable; probably the causes were deep-seated within the earth, changes of pressure and stresses in the semi-plastic under-crust, caused themselves maybe by the great weight of sediments deposited in the artesian basin to the west and on the floor of the Pacific Ocean in the east. Whatever the primary causes, it is known that the elevation did take place, and as might be expected over such a great area, different rock strata yielded in different ways as they were pushed upwards.

In some places the tableland rose as a broad dome; there was a gentle swelling and the margins of the tableland sloped evenly downwards or were warped into broad and simple folds. An example of such folding may be seen in southern New South Wales, where the once flat-bedded Permian sandstones form the surface of the land and slope gently to the coast beneath Jervis Bay, rising again slightly to the east to form the cliffs at Point Perpendicular (see Figure 19).

Although the sandstone surface undoubtedly demonstrated folding, the inclined ramp is not directly due to the earth movement because the shales of Nowra Hill show that overlying weaker rocks have been stripped off by erosion. This illustrates a recurrent problem in attempting to understand the topography of the Eastern Highlands. This is the difficulty of ascertaining to what extent it is a direct product of earth movement—tectonic relief as it is called—and to what extent it is only indirectly so through the influence the arrangement of the rocks by earth movement has on erosion.

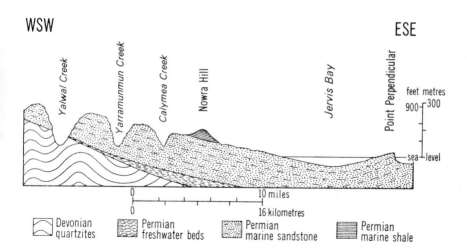

19. Section of edge of Southern Tablelands at Jervis Bay, New South Wales, showing how Permian sandstone has been gently folded and finally submerged to form the bay.

Elsewhere the tableland rose *en masse,* its margins fractured in nearly vertical faults, giving rise to steep escarpments, particularly on its eastern side. Whole areas of country reacted independently and were lifted high above or remained below the general level. Most geologists agree that the Bellenden Ker Range is one solid block which was pushed up as a horst well above the adjoining tableland. Other blocks either did not rise at all or in the general upheaval sank back again even below their former level. Most of the Queensland corridors are probably rift valleys formed in this way, but they must be distinguished from the true river valleys, the deep gorges cut during the intervening time by the streams descending rapidly from the eastern edge of the tableland.

The story of the tableland since its elevation is one of unending erosion and destruction. The rivers were reborn, changing from sluggish, meandering streams to rushing torrents tumbling headlong from the plateau to the sea. At first there must have been innumerable waterfalls, but as the rivers cut their gorges backwards into the hills, the waterfalls gave place to rapids and cataracts. Many waterfalls still exist, generally where some hard stratum of rock is undercut by a softer one, and such usually occur towards the head of the gorges rather than on the edge of the tableland. Most of the Queensland waterfalls occur where beds of the hard resistant lava rhyolite replace the easily decomposed basalts, but the famous Barron Falls are in hard beds of very ancient schist.

Most of the rivers on the eastern side of the tableland are short and flow directly to the sea, their estuaries only navigable where they cross the narrow coastal plains. Some of them follow the pre-existing corridors, but these are mainly occupied by minor streams and north-south tributaries of the larger rivers. A few rivers are longer, particularly where the tableland is low. The Burdekin and Fitzroy rivers in Queensland and the Hunter River in New South Wales rise well to the west, where the main divide is often little more than a mere ridge. This main divide is indeed nearly always on the western side of the main tableland, and the highest mountains are entirely drained by eastward-flowing streams. Apparently the tableland rose so slowly that many rivers were able to cut their beds down in pace with the elevation. Now they run in their middle courses in deep gorges right through the highest part of the tableland.

Contours have been softened by erosion in the intervening ages; escarpments formed by faults have been worn back and their slopes lessened; the shape of horsts has been rounded by the passage of time. Many different geological formations have given rise to much diversity of scenery; the hard rocks have remained as ridges; the softer rocks have been worn down and form the valley floors. At a later stage great floods of basaltic lava again modified the scenery, filling in older valleys on the tableland and diverting many rivers from their courses. New streams are in turn dissecting these lava-flows and in places again exposing the older river gravels far below. All is in a stage of gradual dissolution.

LAKE GEORGE AND LAKE BATHURST

In Queensland and New South Wales most of the spectacular scenery lies on the eastern edge of the tableland and takes the form of precipitous escarpments, deep gorges and waterfalls. Much of the summit area is undulating, the hills and valleys being survivals of a much older landscape. Here and there, nevertheless, are topographical features, not spectacular in themselves but with their own peculiar interest. Among these is Lake George in New South Wales, best described as being sometimes a lake and sometimes not. The chief interest of Lake George is that it is a sunken land or *Senkungsfeld*. It has been thought to be still subsiding, but the uniform level of the high former lake shorelines around it shows there has been no significant earth movement since they formed.

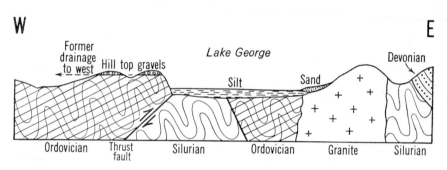

20. Section across Lake George showing associated faulting.

Lake George lies about 25 miles south-west of the city of Goulburn, alongside the main highway to Canberra. A road from Bungendore approaches its southern end. The depressed area, of which the lake occupies the lowest part, is more than 30 miles long and about 12 miles wide. It lies beneath the Cullarin Range to the west. The Great Divide between Tasman and Murray drainage divides here to pass along the Cullarin Range and also along the divide to the east of Lake George, for the latter is an internal drainage basin. The Cullarin Range is not a range in the true sense. From the western side it appears merely as a low ridge sloping gradually to the general level of the tableland near Yass. From the east it is more imposing, as it forms a steep, unbroken escarpment up to 500 feet high. The escarpment is the key to the existence of the lake, for it lies along a fault.

The basin of the *Senkungsfeld* is bow-shaped, the straight bowstring the escarpment on the west, the bow itself a clearly discernible ridge forming the divide between a few small creeks flowing westwards to the lake and other creeks flowing eastwards into the gorge of the Shoalhaven River. When the lake contains water it may be of considerable size, up to 15 miles long and six miles wide.

Since it was first discovered the lake has had many vicissitudes. When dry it is a large saline plain where grass and an introduced salt bush provide fodder for many sheep. Fortunately for farmers it is not liable to sudden flooding. There are more than 100 feet of silt below the surface, and even heavy local rain is either soon evaporated or absorbed into the porous silt. There have been times, nevertheless, when surface water has remained so long that its permanency has been assumed, launches and rowing boats have plied upon its surface and jetties have been built upon the shore. Records show that it was a lake from 1816 to 1830; then it was dry except for short intervals in 1852 and in 1864 until 1874, when it filled again and became a lake until 1900. Later records are not clear, but it was dry in 1925, filled again, and was again dry until January 1949, when it again filled. The record rain of 1950 set seal to this and mainly good years have followed so that the lake has never dried up since then, though during a deep drought in 1967 it was well on the way to doing so.

The noted geographer, T. Griffith Taylor[10] considered the remarkably straight scarp west of Lake George to be a Pleistocene fault scarp with the lake basin dropped down to the east, dismembering a river system which formerly flowed to the Yass River. He thought the fault was still moving because there are rather frequent minor earthquakes, most severe at nearby Gunning where walls have occasionally been cracked and china shaken from the shelves.

However the only fault proven along the scarp so far is a very old one, thrusting up the Cullarin Range from the west. Study of the earthquakes has shown that they relate to a different faultline running north-west through Gunning. Moreover radiocarbon dating of charcoal from emerged lake shorelines show that the lake is older than Griffith Taylor allowed. It is possible that the story of Lake George is more complex than he thought. After the old thrust faulting, the original fault scarp probably was destroyed as erosion reduced the area to a lowlying erosion surface. Then uplift brought about renewed erosion which recreated the basin by etching out weaker rocks east of the fault. A faultline erosion scarp brought character to the landscape once more. But there may well have been renewed faulting more or less along the same line not long ago geologically. Nature usually has fashioned its scenes through complex, rather than simple evolution.

About 12 miles east from Lake George is another lake which also sometimes contains water and sometimes is dry. Lake Bathurst is much smaller and occupies only about five square miles, but it is of interest because it owes it origin to a different and rather unusual cause. Running parallel to the railway near the township of Tarago is Mulwaree Creek, not a large stream, but in wet seasons flowing quite freely. In phases of more effective precipitation in late Pleistocene and Recent time, it probably had a much greater volume than now. At that time there entered Mulwaree Creek from the east, a lesser stream which had a very slight fall. In the course of time the larger

10 *Proceedings Linnean Society, New South Wales,* 1907.

stream deposited gravel at the side of its banks and dammed back the water of its tributary to form a small lake. In intervening periods of less effective precipitation, there has never been sufficient force of water to cut away the barrier, though once or twice the lake has risen sufficiently to overflow slightly. However Mulwaree Creek has managed to cut down the floor of its own valley a little, leaving in the tributary valley a plug of sand and gravel as a terrace above its present floodplain. The catchment area of the lake is not large and it is often dry, but there have been times when boat races have been staged upon its surface.

THE ROOF OF AUSTRALIA

In the south-eastern corner of Australia is the highest and in one way the newest land in the continent. It is undoubtedly the highest, and it is the newest because its very height makes erosion more rapid than in lower places. The process of erosion may be so slow that it is imperceptible within a lifetime, but it goes on inexorably from day to day, year to year and millennium to millennium. The higher the land the more potent is the action of frost and other agents of destruction, and the more rapid the removal of the worn material. At each stage in the lowering of the surface a new land may be said to be born, a land drifting in its contours, however slightly, from that which has gone before. In the course of time the whole landscape is subject to change.

The high lands of south-eastern Australia have been termed the Australian Alps, but this is a misnomer, inasmuch as they are in no sense alpine ranges as this term is usually understood. Alpine mountains are recently folded mountains, but these are essentially part of a tableland, a continuation of the same tableland dealt with in the earlier part of this chapter.

The vertical elevation of the tableland here attained its maximum. The highest portion is from 6,000 to over 7,000 feet above sea level, but it was probably at one time considerably higher. It has already been subjected to continuous erosion for more than a million years and must have been lowered to a certain extent. As it is now, its summit is a continuous wide ridge in the shape of a great bow about 75 miles long. At the northern end of the bow is Mt Kosciusko, 7,316 feet; then the ridge crosses the Victorian border and swings to the west round the headwaters of the Indi or Upper Murray and the Mitta Mitta rivers to Mt Feathertop, 6,307 feet, finally turning to the north again to Victoria's highest point, Mt Bogong, 6,516 feet.

On all sides the tableland originally sloped down more or less evenly to lesser heights. The greatest elevation was over such a limited area that it could hardly have taken place without serious dislocation of the rock masses and the production of extensive faults and fault escarpments. Just where these faults lay is not always easy to trace. Original fault scarps have been worn and obscured, the geological structure of the whole area is very complex, and the numerous rivers have dissected the margins of the high land into a complicated pattern of gorges and valleys, with many separate plateaux, ridges and individual peaks between.

Farther to the west the tableland continues at a lower altitude well into western Victoria, but these western highlands are cut off from those in the east by a curious low-lying gap north of Melbourne. The narrowest part of this gap is near Kilmore about 35 miles from Melbourne, and it forms a natural gateway for both the main railway and road. It is thought to be a northerly continuation of a great rift valley which sank after the main elevation and within which lies the submerged area of Port Phillip.

The Australian Alps have become both a summer and a winter playground for Australians, even though the inner core is surrounded by a tangled mass of lower mountains covered by vast forests of gigantic trees. Some rivers such as the Mitta Mitta, the Ovens and the Tumut on the northern side and the Thomson and Latrobe on the southern, wind through fertile valleys deep within the hills, but there are others such as the Lower Snowy that rush through deep precipitous gorges in lonely country. However many of the construction roads of the Snowy Mountains hydro-electric and irrigation scheme and of the Kiewa scheme in eastern Victoria have been maintained subsequently as tourist roads, and there are also roads specially constructed to reach ski fields.

It is only from a few vantage points that the high inner plateau can be glimpsed in all its grandeur. From Albury on the New South Wales border on clear days in the winter an undulating line of white may be seen away to the south and above the summits of the intervening hills. This is the mass of Mt Bogong over 50 miles away.

A nearer view may be had from the summits of the hills about Bright. Bright is a beautifully situated township nestling in a narrow valley near the headwaters of the Ovens River. It is in the heart of the timber country, and dense forests clothe the surrounding hills and come right down to the out-skirts of the settlement. From the high hill to the east of the town Bogong is clearly visible at a distance of less than 25 miles, its long, otherwise even skyline broken by a few rounded eminences, the contour of a typical table-land. Bright is in the centre of a mass of mountains, for to the south-east and at about the same distance is Mt Hotham, 6,108 feet high, and only 10 miles to the west the bulk of Mt Buffalo, isolated from the main tableland but still 5,654 feet high.

These three mountains, though presenting some differences in structure, exemplify the topography of the high plateau. There are no outstanding peaks, and when the summits are reached it is difficult to determine just which is the highest point. The summits are plains, that of Bogong of great extent, not quite flat, for more resistant masses of rock rise as hummocks above the general level, and shallow gullies radiate outwards, becoming steeper as they near the edge. Between the rocky outcrops there are wide flat expanses covered with grass, often swampy, and in summer gay with many flowers. Snow covers the ground during the whole of the winter months to a depth of many feet, thinner on the hill-tops but accumulating in thick drifts in the hollows and higher valleys. It may persist until September or October, when its final melting turns the mountain streams to raging torrents and swells to flood level the many rivers which here have their source. The summer is mainly mild and clear, with warm sunny days and cold starlit nights. But fierce winds often sweep over the shelterless plains, and snow storms and blizzards may occur even in the middle of summer.

There are no trees on the summits, and the tree line is well below the edge of the higher tableland. At about 5,500 feet a few trees may be seen— stunted, twisted and gnarled snow gums (*Eucalyptus niphophila*), singly or in small groups, clinging tenaciously to life in the more sheltered positions. The snow gums become thicker as the slopes are descended, and from 5,000 to 4,000 feet their smooth white trunks form a continuous forest. At 4,000 feet the great forests begin to appear, a belt of woodland completely surrounding the higher tableland, covering hill and valley alike in an endless mantle of green upwards of 200 feet above the ground. Here are most of the biggest trees in Australia, the mountain and alpine ashes (*E. regnans* and *E. delegatensis*), the candlebark (*E. rubida*), giant stringybark and many others.

The summit of Mt Bogong, like most of the higher parts, is composed of granite in intrusive masses which at one time or another pushed up through the surrounding Ordovician slates and schists. The granites here have been resistant to erosion, and have remained after the softer slates and schists have disappeared or been worn to lower levels. On the Bogong upland plain the granite, where it outcrops, has weathered smoothly into rounded knolls, but on the margin of the tableland it has splintered into sharp jagged ridges which line the sides of the precipitous slopes leading down into deep gorges.

A little farther to the south Mt Hotham and Mt Feathertop have retained their height from rather a different cause. There is no granite here. The rocks are Ordovician slates, but the summits are covered with sheets of hard basalt, residuals of the great lava-flows which covered the land before its elevation. The summits have a similar topography to that of Mt Bogong, except that the terrain is smoother and the slopes into the valleys are more even: ski enthusiasts from Melbourne and beyond find them ideal for their sport.

Mt Buffalo is an isolated plateau west of the main ridge. It is not so high but is noted for its spectacular scenery. It is composed of granite, flattened on top, with steep and often precipitous sides. The granite here has weathered rather differently from that of Mt Bogong and is much more rugged. There are few alpine meadows on the summit and the surface is nearly all bare rock, not in smooth pavements but in serrated ridges and hummocks. Vertical joints have caused the separation of many small hills of well rounded form, and to one hill composed of serried ranks of granite monoliths the name of "The Cathedral" has been given. The rounded forms of the granite here are the result of subsurface rotting of the rock, followed by stripping of granular products from between the sounder boulders and projections. There is little sign of the splintering and shattering due to frost action, which is to be seen on the summit of Mt Townsend in the Snowy Mountains.

The margins of Mt Buffalo are steep and precipitous. Streams have cut downwards along the main lines of jointing, and they tumble in picturesque falls into narrow gorges with vertical walls. Everywhere are precipices or slopes of jagged rock, with great boulders piled in heaps or perched precariously on the cliff edges. The whole is a scene of grandeur and indescribable chaos.

It is probable that this weakness in the granite has led directly to the lower height of Mt Buffalo, compared with its neighbours Bogong and Feathertop. Its very roughness suggests that it is being lowered by erosion at a more rapid rate; on the other hand it must not be overlooked that it may never have been so high, and that it is a separate block mountain bounded by faults.

At the northern end of the arc of high tableland is the highest part of all, the Kosciusko Plateau, its summit over 7,000 feet above sea level. In a way it is the hardest to view as a whole, for it is surrounded on the east, south and north by a maze of deep gorges and rugged ranges, thickly timbered and themselves difficult of access. Approach by road is from the east, and though the slope is steep and long there is little suggestion of height even when the tree line is passed. It is on the western side that the plateau is most imposing, for here it rises like a great wall many thousands of feet above the valley of the upper Murray River. The wall is not vertical but extremely steep, and precipices hundreds of feet high are dwarfed among the forested ridges which radiate downwards.

The top of the plateau is a wide plain not nearly as flat as distant views suggest. Count Strzelecki, the noted explorer and scientist, struggled to the summit in 1840. He found it a dreary region with numerous rocky hillocks rising above the general level. Between the hillocks were extensive meadows covered with grass and alpine plants, and wide hollows of bog and marsh, the source of innumerable rivulets which meander across the summit before plunging downward to join the rivers far below. Strzelecki selected what seemed the highest point and named it Mt Kosciusko in honour of the great Polish patriot. However it seems likely that the summit Strzelecki actually climbed from the west was in fact that lower peak now called Mt Townsend of 7,249 feet. It is also possible that the highest point had previously been surmounted by Dr Lhotsky, a Polish compatriot of Count Strzelecki's and by local stockmen, who were grazing stock nearby before 1840.

The Kosciusko Tableland is of granite, like so many of the other high places, a compact hard granite which has resisted erosion and enabled it to retain its height. There is little difference in the topography compared with that of Mt Bogong, except that here is evidence of the action of glaciers during the Great Ice Age.

The action of ice in the Snowy Mountains was on a much smaller scale than in Tasmania. About 20 square miles from Kosciusko to Mt Twynam was affected. Small individual glaciers formed at the heads of the valleys and cut ampitheatre-like basins with steep walls called cirques. The longest glacier was that of the upper Snowy River valley and was about 5 miles long. There were some hundreds of feet of ice in the cirques sufficient to grind and polish rocky outcrops in its path. This and the presence of moraines and erratics of rocks foreign to their present locality are evidence of the passing of ice.

Apart from moraines and cirques a tangible scenic reminder of the glaciers is the nest of small lakes just beneath the brink of the plateau and right at the source of the Snowy River just to the east of the Charlotte Pass. The beautiful little Blue Lake lies in a cirque, in a rock basin left by the melting ice. The smaller Hedley's Tarn lies just below the Blue Lake on the same stream and is dammed by an old moraine. Down Rawson Pass along the valley to the south-south-west is the moraine-dammed Lake Cootapatamba, 6,740 feet above sea level, and in the valley below it are the remains of two former lakes, their dams of morainal material breached and their waters drained.

All these small cirque and valley glaciers responsible for these attractive modifications of the scenery belonged to one glaciation, although stages in its recession can be recognized. Radiocarbon dates of peat, wood and humic soils related to deposits during this cold period suggest the glaciation took place between 30,000 and 15,000 years ago. There may well have been earlier cirque glaciers of a previous glaciation but any direct evidence of it has been destroyed by the later larger ice masses.

Much study of the highest mountains of Victoria has failed to prove the existence of any glaciers there, though it has shown other effects of colder climate than the present one.

An interesting feature of the topography of the southern tableland is the intricate nature of the river systems. Many rivers are born here. Two great rivers, the Murray and the Murrumbidgee, flow inland to the north and then westward on their long journey to the sea, and there are also many shorter rivers which flow directly southwards through Victoria to Bass Strait. In central and western Victoria the main divide is clear and comparatively simple, running east and west and following generally the highest part of the tableland. East and north of Mt Kosciusko it is extremely complicated and at first sight bewildering.

Geologists differ in their opinions as to whether the river valleys are rift valleys or whether they are due purely to erosion. Both opinions may be partly correct, but erosion has certainly played a tremendous part, even if it did not always determine the first direction of the streams. As in Queensland a clue may be found in the "grain" of the country. Most of the sedimentary rocks in the area are Ordovician slates, overlain by Silurian and then by Devonian strata. All are greatly folded and the axes of the folds run generally in a north-south direction. Where faulting may occur on the eastern margin of the main tableland it might be expected to have the same general trend. The rivers also have this trend, and where the rocks are softer they have excavated their beds to form a series of nearly parallel valleys all running north and south. The peculiar feature of the drainage is that the water does not always flow the same way, and in adjacent valleys the rivers may be flowing in opposite directions. Possibly owing to the presence of more resistant rocks across the "grain" of the country a river may flow one way in a valley for a considerable distance and then double upon itself to flow in a parallel gorge in the opposite direction.

A good example of this is seen in the mountains just west of the Australian capital, Canberra. Here two rivers, the Cotter and the Goodradigbee, flow north in parallel valleys to join the Murrumbidgee. The Murrumbidgee itself rises farther to the west and actually flows south for nearly 80 miles, then near Cooma it turns upon itself and flows north, and finally west, thus completely encircling its tributary streams.

Over still another divide to the west is the Eucumbene River, flowing in another parallel valley south to join the Snowy River, which continues southwards through Victoria to the sea. Then again westwards is another valley with this time a northward-flowing stream, the Tumut River, another tributary of the Murrumbidgee. Westwards again from the Tumut is the divide between the Murrumbidgee and the Murray River basins, and though some of the streams run southward for a while, it is to join others with a westerly trend, all running eventually into the upper portion of the Murray.

This complication of river systems presents a problem to even the most experienced bushman, and has made this portion of the continent very difficult to explore and survey. Nevertheless it is more than a geographical curiosity, for out of it has developed a scheme of vital importance in the national economy, the Snowy Mountains hydro-electric scheme. The natural drainage has been exploited and changed completely in this. Huge reservoirs now cover highlying broad valley plains of the upper Murrumbidgee, Eucumbene and Snowy Rivers on the eastern side. Tunnels pierce the divides and take this water to the much lower valleys of the Murray headwaters and the Tumut on the western side. It is dropped to power stations and then is available for irrigation of the alluvial plains in dry country to the west. As a result of all this, the lower Snowy is deprived of its drainage from the high ranges but the lower Murrumbidgee recovers much of its water through its tributary, the Tumut.

THE GRAMPIAN MOUNTAINS

Last outposts of the Eastern Highlands are the Grampian Mountains in western Victoria. On any small-scale physical map the high land near the coast after turning to the west appears as a belt of varying width from the vicinity of Mt Kosciusko, and forms a continuous backbone in the middle of Victoria nearly to the South Australian border. North of Melbourne is the Kilmore Gap, where the land is lower; thence to Ballarat there is a confused tangle of hills rising from a not very high tableland, and finally the abrupt slopes of the Grampians extending in a rugged north-south arc, the last rocky bastion, overlooking afar the great plains to the south, west and north.

From the summits of almost any of the extinct volcanoes which dot the western plains of Victoria the northern view is of a treeless plain rising so imperceptibly that to the eye it appears quite level. Beyond this is an uneven skyline, the horizon broken by a number of irregular forest-clad mountains.

The Wonderland, Grampian Mountains, Victoria. River gorge and weathered sandstone. *Victorian Railways.*

King River gorge from the railway in wet western Tasmania, with its rain forests.

In winter, but not in every season, the highest may glisten white for several months where the snow lies unmelted, or they may be completely hidden behind flanks of fleecy clouds.

From the northern plains which fall gradually to the Murray River the view is similar and the Grampians are visible from at least forty miles. From here they were first seen by Sir Thomas Mitchell, who wrote in his diary on 11th July 1836:

> "From a high forest hill I first obtained a complete view of a noble range of mountains, rising in the south to a stupendous height, and presenting as bold and picturesque an outline as ever painter imagined. The highest and most eastern summit was hid in the clouds, although the evening was serene."

Mitchell soon after climbed the highest peak, which he named Mt William after the English King. As it was the depth of winter he and his party suffered extremely from the cold. One of the things he noticed was a small species of eucalypt growing right on the summit. This was afterwards named *Eucalyptus alpina* and so far has not been found in any other locality.

Mitchell's description of mountains rising to a stupendous height is perhaps an overstatement, for the surveyed height of Mt William is but 3,829 feet, and there are few other heights above the 3,000 feet level. Nevertheless the Grampians rise so suddenly from regions of low relief that to explorers with eyes long attuned to the even horizon of the plains they must have been imposing indeed. Scenically they are one of the wonders of Victoria.

The general trend of the mountains is north and south, and they lie in a great arc nearly 90 miles long, the convex side of the arc to the east. Actually there is not one main ridge but three, the longest one in the centre. This is the Erra Range, its southern portion very narrow, and pushing out for more than 20 miles into the plain, to terminate in the precipitous bluffs of Mt Abrupt and Mt Sturgeon. To the east of the Erra Range and separated by the gorge of the Wannon River is a shorter and higher ridge on which stands Mt William. This ridge looks down to the east over the fertile valley of the Hopkins River and the town of Ararat. To the west of the Erra Range is the Victoria Gap, a valley less than 700 feet above sea level, and across this the Victoria Range, rising in only one or two places to 2,000 feet. To the west again beyond the Glenelg River is another small ridge, the Black Range, 1,200 feet above sea level. In spite of this, the highlands of eastern Australia may be said to terminate with the Victoria Range. Beyond this is the beginning, scenically, of quite a different world.

Structurally the Grampians are composed mainly of a hard fine to coarse sandstone, white, grey, brown, red or purple. The sandstone, though parted here and there by thin beds of shale, is exceedingly resistant to erosion, and it composes the great bluffs and precipitous escarpments which everywhere cap the mountains. The strata have been determined as Upper Devonian in age, and beneath them is a pavement of older rocks, Ordovician slates and shales with intrusions of grey granite. These form the lower and more fertile portions of the tableland to the east. The older Ordovician rocks in past ages have been intensely folded, but the Upper Devonian rocks have also been disturbed, and instead of lying horizontally they are tilted to the west at an average angle of 30 degrees but occasionally of as much as 60 degrees. This gives a characteristic shape to the mountains, for the slope is generally more gentle to the west, while the main escarpments of such mountains as Mt Abrupt and Mt Sturgeon face the east.

The Grampians lie in a belt of moderate rainfall. There is no tangled rain forest within their deep and secluded gorges nor are there any of the giant gum forests of the ranges farther east. Nevertheless in spite of, or perhaps because of this, the region is a botanist's paradise. Here on the poor sandstone soil, as on similar locations on the Blue Mountains, is a wealth of wild flowers, dainty *Epacris, Boronia, Eriostemon* and hundreds of others, while in the gullies tree ferns, maiden hair and other ferns, though not so abundant as in the east, are still luxuriant. In all over 700 species of plants have been listed from the area.

The exact origin of the Grampians is still in doubt. They are an extension of the highlands to the east and yet they do not quite belong to them. Just when they were elevated to their present height is not quite certain. It is quite probable that there was high land here very early in the Tertiary Period, when the hinterland of eastern Australia was still nearly a plain at a low elevation. In early Tertiary times, when so much of southern Australia sank below sea level, the Grampians remained as land. The Miocene Sea covered the plain to the south and the base of the Grampians was the shore of the sea. The sea lay also to the west, and to the north it filled what is now the basin of the Lower Murray River. The Grampians then jutted out as a high rocky peninsula from other land to the east. In the great uplift of the Tertiary Period they probably experienced further elevation in common with the surrounding sea bed, which then rose some hundreds of feet above sea level.

Erosion is the chief agent which has shaped these mountains through a period which can only be surmised. Only the hardest rocks have remained as relics of still higher land. The slates have disintegrated and been washed away, the apparently hard granites have decomposed, all the land has been lowered; but the tough, resistant sandstones still remain as lofty and extensive ranges. So they may remain for many ages to come, until in the course of time they too will inevitably be fretted away.

Most of the ridges are due to erosion etching out the weak shales from between resistant sandstones, giving rise to eastward facing scarps and gentler western dipslopes. However the easternmost scarp of all is related to a fault. This fault raised the land to the east higher than the Grampians but, subsequently, erosion levelled the whole area exposing weaker underlying rocks east of the fault. Renewed uplift and erosion then removed these weaker rocks to create a new sharp, a faultline erosion scarp facing east. This is the opposite direction to that in which the original fault scarp faced; such are the vicissitudes of landscape history.

The Ice Age in Tasmania

Tasmania is not only the smallest Australian State, but it is also the most diverse in scenery and in climate, and Tasmanians will certainly say, the most beautiful. Within its small area are packed a great plateau, tangled masses of mountains, great forests, innumerable lakes, picturesque waterfalls, fertile valleys and probably the roughest and most inaccessible country in all Australia. The climate is temperate, with a warm summer and a moderately cold winter, but there is a great difference in rainfall between the settled east and the wild west and south-west coast. The east and north coasts have a moderate rainfall of up to 40 inches, the central tableland is drier; but within a few miles, in the western belt, the rain is at times practically continuous, and averages over 100 inches in the year.

My first glimpse of Tasmania was as long ago as 1909, when I was on the staff of the Technological Museum, Sydney, as Collector, and my chief, the late Mr R. T. Baker, was engaged in research on Australian pines. The primary object of my journey to Tasmania was to collect material for this research, a journey which took me into wild and sparsely inhabited districts. On this occasion I did not reach the central plateau, but circled it on its eastern, northern and western sides, eventually to reach the small settlement of Williamsford, in the heart of the mining area of the west coast. At that time there was no road across the island, and going to Williamsford, within 100 miles from Hobart as the crow flies, necessitated a journey north to Launceston, then west along the coastal plain to Burnie, then by narrow-gauge railway south to Zeehan, and again 20 miles to the east by another small railway.

The time of this visit was in the depth of winter.

The single railway line from Burnie ran southward for a few miles through rich farm land with a moderate but adequate rainfall. Soon it began to wind through hilly forest country, becoming rougher and rougher with every mile, the hills higher and higher, the valleys deeper and more precipitous. Dark clouds obscured the skies ahead, and then the train suddenly ran into the rain. The rapidity of the change was remarkable. As the train swung round the unending curves, occasional glimpses to the north revealed the sun still shining in the regions just left, then the distant light faded. It rained and kept on raining. During the whole of the winter it seldom ceases, and in the summer, though there are dry interludes, it rains more often than not. Returning the same way a few weeks later the train passed from the rain to the sunshine at the same place, and the transition was just as rapid. The rainfall of the west is over 100 inches a year, not in sudden tropical down-pours, but in steady relentless falls, sometimes easing to a heavy mist, but never actually ceasing, day and night, for weeks and months on end.

Zeehan is now a mere ghost town, but 40 years ago it was populous and prosperous, its existence made possible by the rich copper mines in the vicinity. The traveller who stayed overnight was soon made aware of the town's pride, the railway yard. To most newcomers it seemed unpleasant, churned into a muddy quagmire by the traffic to and from the daily train. But, as the local inhabitants pointed out, it was flat, the only flat land in a region where the mountains stand on end and one-in-two slopes predominate.

From Zeehan a narrow-gauge railway ran east for about 20 miles through a winding valley to Williamsford, an even smaller town, depending for its existence on the copper mine at Mt Reid, some five miles away and about 2,000 feet higher up. Since then, the ores have been worked out, the mine, town and railway no longer exist. At the time of this, my first visit, there was a strike at the mine, and the train was running once a week, so that, rather than wait another three days at Zeehan, I put some lunch in my pocket and decided to walk.

The best way to see any country intimately is to walk through it. With a whole day ahead and only 20 miles to cover there was leisure to look around. It is on such a walk that the wildness of western Tasmania is realized. This part, with its railway and scattered town, was thickly settled compared to the wilder areas beyond, yet in 20 miles not a human being was seen, nor was there one human habitation. All about was the virgin bush, untouched save for the narrow clearing where the ribbons of steel wound in and out around the side of the valley.

The day was propitious, for even the rain cleared at times. Then swathes of cloud rolled down and partly filled the valley, so that the hill-tops were hidden from view. Every few yards little runnels of water rippled from the hills, and once the track crossed a larger stream on a viaduct at the foot of a waterfall that fell spectacularly over a lofty cliff. Leafless forests of beeches clothed the hillsides, their hoary trunks green with moss and lichens, and between them giant eucalypts towered overhead. Here for the first time was the ubiquitous "horizontal", binding the forest into an impenetrable fastness. Ferns were everywhere, giant tree ferns in every little gully, and masses of delicate maiden hair beneath every tiny ledge. Moisture dripped from every leaf and twig.

Some indentations on a piece of sandstone attracted attention, and fracture with a geological hammer revealed numerous impressions, recognizable as the casts of lamp shells, organisms that had lived in the ancient seas once flowing over this spot. To geologists who have worked out the succession of the formations in this part of the world it is known that this sea was in the Ordovician Period, some 450 million years before our time. There are also older Cambrian and even Pre-Cambrian rocks.

Here then is a fitting starting-point for the scenery of Tasmania, for here literally are the foundations of this ancient land. Fossil shells and trilobites have been found in many parts of western Tasmania, as well as fossils of an even earlier and again of a later period. In the later Silurian Period there were even coral reefs in this part of the world, similar in general structure to the coral reefs now living on tropical coasts, but composed of corals long since extinct.

It is in solitudes like these that the best setting is found for the contemplation of things linking the remote past with the landscape of the present day. Buried in these rocks is evidence of the changes that have taken place not once, but many times, not only in the geography, but in life itself.

On another occasion, westward from a mountain peak, a distant glimpse of the sea evoked another picture, a picture of the mountainous land that once stretched far over what is now the ocean bed. In the chapter on South Australia reference was made to the Permian Ice Age and to the mysterious continent of Gondwanaland. That land lay to the west of Tasmania in Permian times is certain, for there is evidence in many places that land ice from the Permian mountains overrode Tasmania from the west. A Permian sea also to the east, for Permian marine rocks are found on the eastern side of Tasmania. The western part therefore was then land, and we may look on this sea as a fragment of old Gondwanaland, a fragment that remained when the greater part drifted away. We may go a little further, and say that western Tasmania is another old land surface, a part never submerged beneath the sea since the middle of the Devonian Period, about 380 million years ago.

In the wet gorges about Williamsford the view is restricted to the immediate surroundings, and though there is much interesting detail, it needs a broader outlook to visualize the general landscape. It is on the great tableland only about 25 miles away to the east, but separated nevertheless by some of the highest peaks in the island, that such a landscape may be found. This is an entirely different region with an entirely different climate. While the rain drips endlessly at Williamsford, it is almost certain that the sun is shining brightly these few short miles away. The rainfall on the tableland is low, little more than 20 inches a year, but the mountains on its western border often rise to 5,000 feet or more and are covered with snow for the whole of the winter. From them comes an inexhaustible supply of water to feed the rivers and fill the lakes, and make this probably the best watered region in Australia.

The great central tableland covers about two-thirds of Tasmania, and its northern portion now averages about 3,000 feet above sea level. During the Pleistocene Epoch glaciers eroded it a great deal yet unequally, excavating lake basins but rounding the summits rather than grinding them down substantially. The Walls of Jerusalem projected above the general plateau surface before the Ice Age. It is not an even tableland but is tilted, so that its northern edge is the highest, while to the south it slopes gradually downwards, to disappear beneath the ocean in a long ridge many hundreds of miles to the south.

On the northern edge of the tableland are the Western Tiers, forming a mighty wall that rises abruptly from the coastal belt and is visible from far out to sea. From the north the Western Tiers present an almost even skyline nearly 4,000 feet above sea level, but some parts rise even higher. Ironstone Mountain is 4,736 feet high, and there are a number of other peaks well above 4,000 feet. So straight and unbroken are the Tiers they have the appearance of a fault scarp. However along most of its length there are no faults and it is there simply a very great erosion scarp. However, Bass Strait is in part a rift valley, which sank far below its present level in early Tertiary times, admitting the sea to parts of the north Tasmanian coast, over much of southern and western Victoria, south-western New South Wales, South Australia and the Nullarbor Plain.

The Western Tiers are like the edge of a tilted table, the steep escarpment to the north, the flat top sloping gradually to the south. This is the region of the Tasmanian Lakes, lakes large and small, in all sizes and shapes, from mere ponds a few yards across to others many square miles in extent. On the eastern side is the Great Lake, the largest in Tasmania. This is a magnificent sheet of water over 14 miles from north to south and averaging over five miles in width, its deep deep placid waters reflecting the trees that line its lonely shores. It is not as inaccessible as it used to be, for a metal road now runs along its western shore and connects it to the town of Deloraine below the mountains to the north. Another good road passes east of Great Lake from the hydro-electric works at Poatina on the Tiers, which tap the power of the lake, to join the previous road near its natural outlet at the southern end.

The plateau bordering the Great Lake is at about its highest and is here close to 4,000 feet. The Great Lake itself lies in a wider amphitheatre of low hills open to the south. To the west of the Great Lake is an uninhabited region reaching for over 50 miles to the west coast railway at Zeehan. The plateau extends for about 25 miles. It is a wild and desolate region, under snow for most of the winter, bleak and wind-swept. There are few trees. Those that exist are small eucalypts, bent and gnarled, nestling in narrow belts in the lee of the low hills which surround the innumerable lakes. There are now several roads built for engineering and timber-getting such as the one to Lake Augusta. The lakes vary in size from two miles across, as with Lake Augusta, to a hundred yards or so in diameter. I do not know if they have ever been counted, but there must be nearly 200 within this area.

These many small and medium-sized lakes lie within the former extent of a small ice cap in which ice moved outwards in all directions. The ice eroded hollows in the rock and deposited barriers of moraine to create a landscape closely resembling that of the Barren Grounds of Arctic Canada. The larger lakes to the east and south-east were also thought to be due to ice action. But with the exception of Lake St Clair no evidence of Pleistocene glaciers has been found around them. Their origin is obscure but they may be due to slight northward tilting back of the plateau against the southward flow of the rivers, which had developed in response to the earlier major tilt to the south.

On the verge of the main divide is some of the roughest and highest country in Tasmania. The Great Western Tiers here bend southwards, and the margin of the tableland is deeply dissected by many narrow gorges. The Mersey and Forth rivers are nearly 2,000 feet below the top of the tableland; they flow to the north in narrow parallel gulfs a few miles apart. Beyond them to the west is the remnant of the plateau. Above it rise the spectacular peaks of Cradle Mountain, 5,069 feet, and Barn Bluff, 5,114 feet. A little farther to the south at the head of the gorges is another group of high peaks, the head-waters also of streams running to the west and south. Of these mountains Mt Pelion West is 5,305 feet and Mt Ossa 5,220 feet above sea level.

South of Mt Ossa, and fed by the melting snows from many mountains, is the most beautiful of the Tasmanian lakes, Lake St Clair. Lake St Clair is a

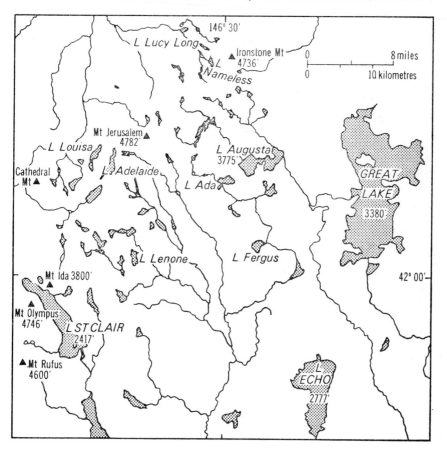

21. Map of western part of the central tableland of Tasmania showing the great number of lakes.

narrow sheet of water over eight miles long and nearly two miles wide, its surface 2,417 feet above the sea, lying in a winding valley overshadowed by lofty tree-clad mountains. On the eastern shore, rising almost sheer from the water, is the Traveller Range, culminating in Mt Ida, while to the west are Mt Manfred, 4,550 feet, Mt Cuvier, Mt Byron and Mt Olympus 4,746 feet. The last literally overshadows the lake, for the trigonometrical station is little more than a mile from the shore and 2,000 feet above it. The highway connecting Hobart with Queenstown passes close to the southern shore of Lake St Clair, to which it is joined by a branch road.

Other nests of glacial lakes are in the south and west, mostly mountain tarns in the higher country. Some are near the summits of the King William Range, which is the divide between the Derwent and the wild Gordon River. Another nest of tarns occurs on the picturesque Frenchman's Cap, 4,739 feet, which, though 30 miles from the sea, was first seen and so named from passing ships.

All the water from the Great Lake, Lake St Clair and the lakes between drains south with the general slope of the tableland into the Derwent River, which flows south-east to reach the sea at Hobart. The Derwent River nearly divides Tasmania into two, and south and west of it lies the roughest and most impenetrable part of Tasmania, probably of Australia. It is a tangled mass of mountain peaks, their sides clothed with beech and eucalypt forest, their valleys mainly swamps and covered either with button grass or dense low scrub. Few penetrate this wilderness, and until recently much of it was unexplored. Those who have fought their way through the scrub have mainly been hardy prospectors seeking the rare metal osmiridium. From the east a track was cut to the headwaters of the Gordon River, where the little hamlet of Adamsfield was established as the headquarters of the osmiridium miners. But ball-point pens do not require osmiridium as did fountain-pen nibs so the mining has ceased and the hamlet at Adamsfield has gone.

The Gordon River is here the main stream, running generally parallel to the Derwent but in the opposite direction, that is, the north-west. Looking at all but the most recent of official maps, one could still find an area of about 60 miles by 30 miles across which is printed, "Uninhabited country, the valleys covered with impenetrable bauera and horizontal scrub." This country continues to the rugged and stormbeaten south and south-west coasts. In the last few years, roads have been pushed forward into the heart of this country in connection with a very large hydro-electric scheme on the Gordon River. This will drown many of the broad valley bottoms here, including the very beautiful Lake Pedder. Many people are opposed to such developments, thinking that it won't be long before power will be obtained from atomic fission and fusion, and that it is shortsighted to destroy some of the most precious elements of the natural heritage when this lies in the future.

Throughout the centre of Tasmania an interesting link with the geological past is the rare and unique thylacine or Tasmanian wolf, a creature simulat-

Cradle Mountain, Tasmania. The ice swept over the plateau edge, smoothing the flank of the mountain and excavating the basins of Lakes Dove and Wilkes (nearly hidden). *J. N. Jennings, Australian National University.*

Tessellated Pavement, Eaglehawk Neck, Tasmania. Flat shore platform in regularly jointed sandstone due to weathering and wave action. *Tasmanian Tourist and Immigration Department.*

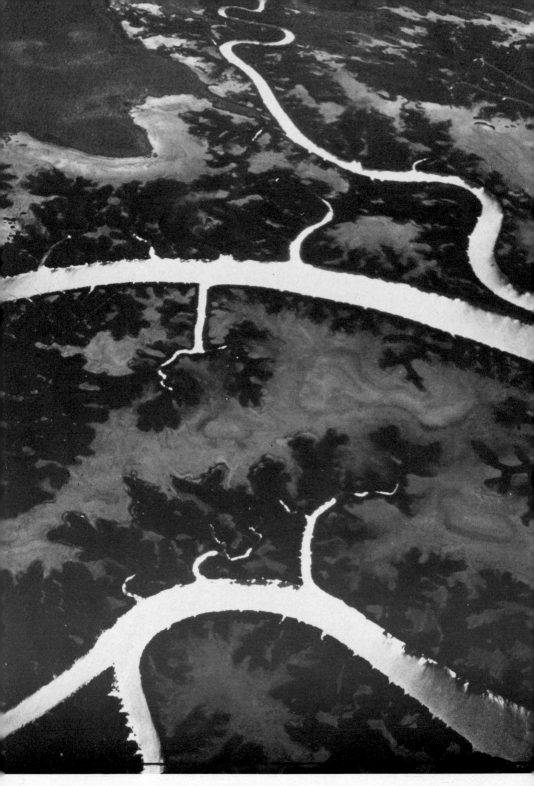

Patterns of estuary channel, mangrove swamp (black), and tidal mud flat (grey), Cambridge Gulf, Western Australia. *J. Cavanagh, CSIRO Land Research*.

ing a wolf in appearance but a true marsupial nevertheless. Bones of the thylacine have been found fossil on the mainland but the animal has never been seen alive there. Some 50 or 60 years ago it was quite common in Tasmania and was a source of trouble to the outlying settler, since it preyed on sheep as well as on the small marsupials which were its natural food. Since the last specimen was caught in 1930 there have been very few and uncertain traces of its existence. It is at least in danger of extinction, probably extinct. This will be the fate of other native species if a real effort is not made to preserve adequate living spaces for them in which they may survive.

A link with an even more remote past is a curious little crustacean like a shrimp found in many of the small mountain tarns. This has aroused world-wide interest, for not only is it a very primitive type, but it has been found fossil in the Permian rocks of Europe. Its name is *Anaspides*.

The extreme south-west of Tasmania is not only inhospitable inland, but its coast is exceedingly dangerous for navigation. In the days of the old sailing ships the prevalent westerly gales made it a lee shore which was a perpetual hazard. How many ships were wrecked there will never be accurately known, for even if survivors did reach the shore their chance of rescue was remote. Fragments of unknown wrecks are even now occasionally discovered on the uninhabited shores. What became of their crews nobody knows, whether they were drowned at sea, or reaching shore, attempted to find their way overland, to be lost in the impenetrable maze of horizontal scrub.

The forces that have shaped the present scenery of Tasmania started their work far back in geological history. In western Tasmania the foundation rocks belong to the very early geological periods; in the centre and on the east they are a good deal younger. In the west there are sandstones of the Cambrian Period, slates and sandstones of the Ordovician Period, and Silurian conglomerates, quartzites and limestones. The Cambrian rocks rest with a great erosional break on thick sequences of Pre-Cambrian rocks, themselves falling into three groups between the deposition of which there were two great periods of mountain-building earth movements. There are also granites and other igneous rocks which penetrated the old sediments during the various upheavals.

As on the mainland of Australia, the older formations have been compressed and folded, contorted and shattered, and thrust one upon the other. Here also the angles at which the strata lie, and their relative hardness and resistance to wear, have largely determined the present shape of the hills and valleys.

The western portion of Tasmania lay under the sea till early Devonian times. The ocean extended far to the west but whether its eastern shore lay within Tasmania is uncertain. The intricate mingling of very early rocks, both hard and soft, lying at all angles to the horizon, has given great variety to the landscape.

Among the strata producing characteristic scenery are the very hard conglomerates laid down in the early part of the Ordovician Period. In the coastal waters and on the coastal plains of those times beds of pure sand were laid down as well as boulder beds of considerable extent. The sand has now hardened to quartzite, and the boulder beds to masses of very coarse conglomerate, in which some individual boulders are several feet in diameter.

South of Queenstown and between it and the spacious waters of Macquarie Harbour is a tangled mass of mountains rising to above 3,000 feet. Many are residuals of hard Silurian rocks, their summits culminating in steep escarpments. The quartzite capping the mountains is often of dazzling whiteness, and from the distance gives the appearance of fields of snow. The escarpments of Ordovician conglomerate are particularly wild and picturesque, forbidding ramparts that are often unscalable. The number of these peaks is many, amongst them Mt Roland and St Valentine's Peak, while others have been called after famous geologists and zoologists, Murchison, Tyndall, Sedgwick, Geikie, Lyell, Owen, Jukes and Darwin.

In a much later period, the Triassic, and in the periods following it, events took place which have had a tremendous effect on the topography of central and eastern Tasmania. In Triassic times freshwater lakes spread over much of the land, and in these a large series of sandstones was laid down.

Such a tremendous catastrophe, even though spread over a long interval of time, would have its repercussion on the neighbouring land. In the middle of Jurassic times about 165 million years ago Tasmania was violently upheaved. Great tongues of liquid material were forced upwards between and through the strata. It is not known if any of this material reached the surface and flowed out as lava, but masses of it forced its way sideways for great distances between the bedding planes of the sandstones. Later it cooled and solidified into a dark-coloured rock of coarse texture known as diabase, or more correctly, dolerite. The sandstones were so affected by the dolerite intrusions that they were sometimes floated entirely upwards, and their remnants now remain like islands embedded in a solidified sea. There were similar happenings in South Africa where there are also great areas of Jurassic dolerite. Many geologists contend that it was at this time that the continent of Gondwanaland began to drift apart.

A curious feature of the dolerite is its almost universal columnar structure. We saw this structure in the Permian lava-flows near Kiama. It produces much spectacular scenery in many parts of the world. It seems to have been caused by contraction cracks at right angles to the surface of lava-flows and other igneous formations, regular, closely packed, hexagonal columns developing as the mass cooled.

Most of the high land of eastern and central Tasmania is covered by sheets of dolerite, including Ben Lomond, the highest peak, and such mountains as Mt Wellington in the south. In the northern parts of the main tableland the dolerite sheets are of enormous extent, and, one above the other, form practically the whole of the plateau. The solidity of the tableland in this part is probably due to the dolerite, for it is exceedingly hard and resistant to erosion, forming a protective capping through which the rivers have not yet succeeded in cutting a passage. It is only on the margins of the tableland, where other formations appear, that great gorges cut into the high land.

Most of the sandstone lying above the dolerite intrusions has long since disappeared.

After the dolerite intrusions there was a long period of quiescence, probably for 70 to 80 million years, and right up to the beginnings of the Tertiary Period. During all this time Tasmania had probably been of no great altitude and was also a part of the mainland. In the early Tertiary Period began the great subsidence that formed the rift valley of Bass Strait and made Tasmania for the first time an island. It was not a permanent island, for before its final severance, it was yet again to be united with the mainland at least once and possibly two or three times.

During the Tertiary Period the land began to rise again and the sea receded to very nearly its present position. This rise included the great "Kosciusko Uplift", and in the previous chapter we saw how it affected the whole east coast of Australia as far as Cape York. During this uplift Tasmania was elevated into a tableland about 5,000 feet high and tilted as we have already seen towards the south.

This brings us to the inception of the Pleistocene Epoch or Great Glacial Age, a period of much significance throughout the world, for it is so closely bound with the advent and development of man himself. In northern Europe and America the Glacial Age has been studied perhaps more than any other, and the succession of phases of intense glaciation with warm climatic interludes has been worked out in great detail.

In the chapter on Port Jackson we have already seen how the level of the sea ebbed and rose during the Great Ice Age, but in Tasmania it was the actual glaciation that had marked effect on the present scenery. With the exception of a small area on the Kosciusko Plateau it is the only part of Australia that was actually buried beneath ice, though, as already stated, other parts had been so buried in more remote ages.

During the last great glaciation to affect Europe and North America, about 1,500 square miles of Tasmania were also buried under ice. This was most continuous from the western part of the central plateau to the great ranges between Cradle Mountain and Mt Olympus. Here were radiating ice caps and glaciers so thick as practically to drown all the relief, with just a peak projecting here and there like the nunataks of Greenland. The effects of the ice cap on the plateau have already been described. Elsewhere in western Tasmania there were smaller glaciers in the mountain valleys and at their heads, which did not bury the divides except here and there where the ice pressed together from different corners. At Gormanston in the west, one such glacier was known to have been in existence 25,000 years ago from radiocarbon dating. When the climate ameliorated towards the end of this cold period, the larger ice masses retreated and smaller valley and cirque glaciers formed within their former greater areas.

The glacial landforms of this glaciation are so young and fresh—some are probably no more than 8-10,000 years old—that not a great deal of change has been made to them by weathering and rivers since the ice melted. Some glacial lakes have been filled in with sediment; whilst others have had their

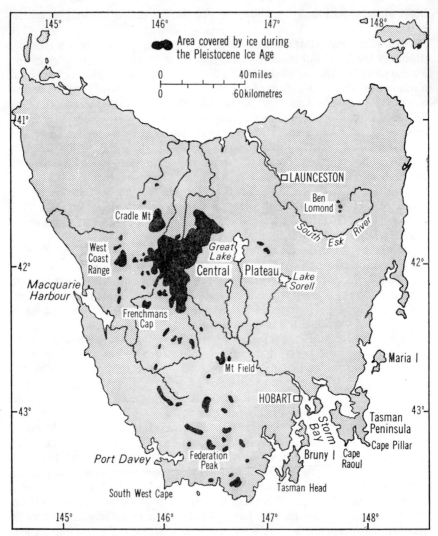

22.   Map of Tasmania showing the area known to have been covered by Pleistocene glacier ice.

moraine barriers breached by river erosion and so have drained. Small gorges have been cut in the floors of valleys eroded previously by ice. But in general the glaciated landscape looks much as when the ice melting uncovered it.

However recently some evidences of an earlier glaciation have been recognized. In the Forth Valley in damsite borings, a loose, fresh glacial moraine of the last glaciation was found on top of a moraine consolidated to the strength of a hard rock or tillite. Similar tillite crops out in the upper Arm valley between the Forth and Mersey valleys. These are thought to belong to an earlier glaciation as are very deeply weathered glacial tills on the central plateau and south of Lake St Clair. In addition there are huge boulders in the Vale of Belvoir north-west of Cradle Mountain foreign to the local rocks. They are however well beyond the limits of the fresh moraines of the main glaciation. These boulders are thought to be glacial erratics of the glaciation to which the very hard and the very weathered moraines belong. It is possible that this earlier glaciation contributed to the erosional work of ice in Tasmania but it is very hard to distinguish its effects.

During both these glaciations and earlier Pleistocene ones of which there is no evidence in Tasmania, Tasmania was joined to the mainland again. The withdrawal of the water from the sea to form enormous ice caps in the northern continents brought sea level below the submarine ridges linking Australia to neighbouring islands. Land was continuous from south of Tasmania to New Guinea. Kangaroo Island was also joined to South Australia and many islands off the Queensland coast were also connected to the mainland. This temporary joining of islands to Australia took place several times with the changing climatic history.

Let us now consider the effects of the smaller glaciers in the valleys, on the sides of ranges and on high individual mountains. These glaciers produced some of the most striking and typical of today's scenery. As they pushed down the sides of the mountains they gouged out broad amphitheatre-like valleys known as cirques, leaving the top of the mountain bow-shaped, with a jagged serrated rim. The inner face of the cirque is often stepped or terraced with cliff faces of the harder rocks which were once great ice falls. Below the cliffs the valley bottom is generally hollowed out by the downward thrust of the ice, and masses of glacial débris (moraines) have been left to make a further barrier to the existing drainage. The hollows are now filled by small glacial lakes, beautiful deep bodies of water, frequently nestling in a whole series down the valley, one below the other.

A perfect example of glacial topography is at the foot of Cradle Mountain in the Scenic Reserve. Cradle Mountain looked at from the west appears as a serrated arc of columnar dolerite, many of the single columns isolated and standing in lofty pillars on the skyline. Farther to the south and also visible from the same viewpoint is the massive block of Barn Bluff, also of columnar dolerite but apparently untouched by glacial action. At the base of Cradle Mountain the ground dips suddenly to a deep gorge, and cutting into the foreground the original course of two great glaciers is clearly visible. Here there must have been spectacular icefalls as the glaciers plunged over the edge and gouged out two deep semicircular valleys. Broad mounds of moraine lie athwart the outlets to the valleys, and behind them nestle the beautiful little blue Crater Lake and Lake Dove, with other similar lakes in the valleys.

Frenchman's Cap, farther to the west, is almost surrounded by cirques, and here too is another nest of typical glacial lakes. In the south, in regions difficult of access, aerial photography and mapping have brought more and more glacial mountains to light. In Mt Anne vertical cliffs rise more than 1,000 feet above the placid waters of Lake Judd, and the Arthur Ranges farther to the south contain many magnificent examples of cirque topography.

Though the Arthur Ranges are only 50 miles in a direct line from Hobart they have been amongst the most difficult to approach. Federation Peak at the south-eastern end of the range is perhaps the most spectacular of all Tasmanian mountains, yet exploring parties only succeeded in reaching its base in 1947. It was finally ascended in January 1949 by a party from the Geelong College Exploration Society led by John Bechervaise.[11] The mountain is marked by a gigantic cirque, with a rugged semi-circular rim open to the south-east and surmounted by lofty pinnacles of quartzite and schist separated by vertical chasms 2,000 feet deep. Right in the centre of the amphitheatre is Lake Geeves, its deep waters actually overhung by the summit more than 2,000 feet above. There is another small glacial lake in a hollow almost on the summit at an elevation of over 4,000 feet, separated from the highest point by a tremendous vertical chasm.

The glacier that excavated the cirque in Federation Peak and reduced the mountain to a mere shell must have been of tremendous power. The freshness of the topography suggests that the glacier belonged to the last phase of glaciation.

The multitude of lakes on the tableland west of the Great Lake are on a nearly flat plain; some are dammed by moraines, some lie in hollows scooped out by the ice. Lake St Clair, on the other hand, is in a deep valley, with dolerite cliffs often rising sheer for hundreds of feet above the water. The water itself is very deep, 530 feet at the maximum. Near the bottom end of the lake there is a drowned moraine and then below the lake there is a great sequence of arcuate moraines. However most of the lake basis is gouged out from the bedrock because a great deal of ice converged on this valley and was capable of much erosion.

It used to be thought that the Pleistocene glaciers reached the west coast and that Port Davey is a glacial fiord. However better examination as a result of improved transport, for instance, by helicopter, and aerial mapping has made it clear that both Port Davey and Macquarie Harbour are simply drowned river valleys. The small glaciers of Arthur Range and Federation Peak deposited their outermost moraines about 15 miles from the head of Port Davey. Larger glaciers reached closer to Macquarie Harbour than this but they still fell short of the coast.

In the last glacial phase Tasmania became joined to the mainland, only to revert to an island once more with the melting of the ice. It is quite possible that during this time the Tasmanian aboriginal migrated across what is now Bass Strait.

There is evidence of aboriginal activity at several sites between 20,000 and 25,000 years ago in the mainland whereas the oldest sites yet dated in Tasmania only reach back to 8-9,000 years ago when the sea had already risen

[11] Australian Geographical Magazine *Walkabout*, April 1949.

sufficiently to drown most of Bass Strait. Nevertheless archaeologists think the Tasmanian aborigines lived on Bass Strait when it was dry land and crossed to Tasmania while it was still a peninsula. Such differences of race and culture as there were between the aborigines of Tasmania and Australia are thought to be due to isolation since the sea became a barrier once more about 12,000 years ago. The Tasmanian people were able to survive this separation but not the coming of the white man with his weapons and his diseases.

There has been little change in Tasmanian topography since the close of the Pleistocene Epoch. Erosion has recommenced its endless cycle, but its effects in the last 10,000 years or so are yet too small to be visible on any large scale. The deposits of moraine, though generally covered with forest growth, are in many places almost as fresh as when they were left by the retreating ice. In the flat-bottomed valleys excavated by the ice, swampy plains are covered with tussocks of button grass, and these are slowly accumulating into deposits of peat. The mountain tops are being further splintered by frost, the rivers are deepening their valleys, the sea is battering and wearing the coast; but the processes are too slow to have been perceptible within the time of our short human record.

The Australian Coast

Let us begin with a map—any map. Maps are amongst the most amazing productions of civilized man. A good map epitomizes on one sheet of paper not only a wealth of documentary information, but it provides the illustrations as well. It shows not only the shape and size of a country but also the height and form of its mountains, its lakes and rivers and plains, its forests and roads, railways and cities. It can be specialized to show the geological formations, the distributions of plants and animals, industries, the incidence of rainfall, magnetic variation, or any other type of information. It can be on such a small scale that a whole continent may be visualized as if viewed from a distance of thousands of miles, or it may be on such a large scale that the smallest detail may be studied. More effort may go into the making of one map than in the writing of a dozen books, and to be understood it does not need translation from a foreign tongue.

In a comparative study of maps of different countries the most obvious thing is the great difference in the shapes of the coastlines. The coasts may be even for great distances, they may be deeply indented by gulfs and bays, they may be bordered by numerous islands, or the land may be intersected by inland seas. Questions that might be asked are, What significance should be given to such differences? Are there any fundamental principles to explain them? Do we really exaggerate their importance? After, all the vertical limits of physical features on the earth's surface are very small in relation to the earth's bulk. If a globe of the earth were prepared to exact scale, the highest mountain and the deepest sea would each be less than one-thousandth of its diameter, and the surface to the eye would appear practically smooth.

To human perceptions these irregularities are nevertheless on a sufficient scale to be impressive and the differences between them well worth study. Thus we can ask why the coast of Australia should be even while that of the South Island of New Zealand is indented by fiords. Why is there no Mediterranean or Baltic Sea within Australia? Why should there be archipelagos off some parts of the coast and not off others? What is the explanation of enclosed bays like Port Phillip, or of long narrowing inlets like Spencer Gulf and St Vincent Gulf? Why should there be a beach in one place 90 miles long, while in others there are no beaches at all?

The questions are not always easy to answer, but there is inevitably some underlying cause even if it is not understood. When studying the physical features of land surfaces, we may find evidence in cuttings and quarries, on the sides of valleys and in the shafts of mines. This evidence ceases at the shoreline, but many techniques have been devised to explore the nature of the sea bed, largely as a result of the search for minerals there in the course of the last few decades. Many kinds of borer have been devised either to take shallow cores of the young sediments on the bottom or for drilling thousands of feet into the bedrocks beneath. The results obtained in these ways are supplemented by various geophysical methods. Electromagnetic waves from "sparker" equipment, natural earthquake waves and artificial earthquake waves from explosions made on the sea floor penetrate to various depths into the submarine sediments and crust. What happens to these waves varies with the thicknesses and dispositions of the various deposits. The waves are reflected and refracted in diverse ways and their behaviour can be recorded. By their study a great deal can be learnt of the structure of the sea bed. As yet, however, not a great deal of the new knowledge acquired in these ways around Australia, particularly in oil search, has been published. It is necessary to rely on deductive reasoning from a limited number of facts which themselves may be interpreted in different ways. With these reservations let us take a map of Australia, paying particular attention to its coastline.

Only two major indentations are visible, the wide shallow one of the Great Australian Bight in the south, and the more enclosed one of the Gulf of Carpentaria in the north. The one large island, omitting New Guinea, is Tasmania, though Kangaroo Island and Melville Island perhaps deserve the title. Tasmania has already been shown to be part of the mainland, separated at first by the subsidence of Bass Strait, and afterwards by fluctuations of the sea during the Pleistocene Epoch. The others will be referred to later in this chapter. What is intended at this stage is to emphasize the compact nature of the whole continent. One other continent, Africa, shows the same compactness and evenness of coastline, and curiously enough Australia and Africa have certain analogies in their geological histories shared by no other continent.

The first principle that can be deduced is that of stability. The coastline of Australia is an old one. For long ages there have been none of the major convulsions of the earth's crust that have resulted in the peculiar contours of many other great land masses. There have been inland seas in past ages, but they have dried up and are now part of the land. There have been deep gulfs penetrating far into the interior, but they too have long since gone. The shape of Australia has remained substantially as it is for a very long period of time. There have been slight changes, local modifications of the coast by subsidence in limited areas, by fluctuations of sea level, and by the action of the sea itself.

Much of the most picturesque detail of the seashore is produced by the erosive power of waves. These in time can modify coastal topography, but it is doubtful if they can ever produce major geographical changes. The action of waves is limited in depth even in the greatest of storms. Waves hurl themselves at the base of cliffs, and with grains of sand and small stones as cutting tools, they gradually eat it back. It is a slow gnawing action and ceases a few feet below low tide level. As the cliffs recede a shallow platform thus widens below the water, and unless there is a change of sea level this platform remains untouched by further erosion. Harder masses of rock often remain entirely separated from the main cliff face, appearing as stacks or rocky islets rising above the eroded platform. As these become further eroded they may become dangerous reefs or bomboras over which the rollers rise and break.

In this way the sea may excavate shallow bays but never deep gulfs going far into the land. The sea may even act as a protective agent by piling sand along the shore in the form of beaches. These act as buffers and protect the shore from more erosion. The water may cut the coast back in one locality and build it up in another.

Material worn by the waves may be sand or mud, with pebbles or even large boulders from the harder rocks. This is dispersed on the neighbouring sea bottom. The finer material is often spread widely; the coarsest material cannot be removed far from the immediate vicinity. Material worn from the shore is augmented by sand and mud brought down by rivers, particularly in flood time. If the neighbouring sea bottom is shallow the rivers themselves may build deltas at their mouths. The deltas of the Nile and the Mississippi are too well known for comment, and that of the Tigris and Euphrates rivers at the head of the Persian Gulf has moved the seashore backwards over 120 miles within historical times.

Australia has few deltas protruding from the coast; the Burdekin River in Queensland has an actively growing delta (Fig. 23) and so has the Herbert River. At the other end of the continent there is the delta of the De Grey River. There are more deltas in estuaries and in lagoons separated from the sea by sand barriers as, for example, in Lake Illawarra and in the Gippsland Lakes. However as a whole the Australian continent is poorly off for deltas. In some sectors the sea is too deep close to the shore to allow rivers to do this construction readily, but there are other more important factors. Because of the comparative lack of indentations, much more of this continent's coast is exposed to powerful ocean waves which distribute and even out the sand and silt brought down to the coast by its rivers, instead of allowing it to accumulate at their mouths. Moreover there is the simple fact that Australia is the driest of the continents apart from Antarctica and into the bargain it is a hot one evaporating much of the rain. Therefore it lacks runoff to gather into rivers which might build out deltas. We need only remember to illustrate this that the U.S.S.R. has five rivers, each one of which discharges more runoff than all of Australia's rivers.

Apart from deltas sandbanks may develop well off shore when the water is shallow; but their shape and distribution are often very changeable and are dependent on variations of tides and ocean currents. Sandbanks, though treacherous to shipping, often break the main force of the waves and protect the land behind. When the sea is deep close to the shore the worn detritus

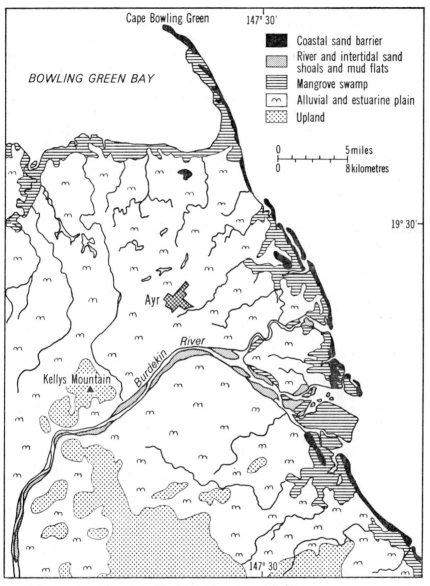

23. The growing delta of the Burdekin River, Queensland.

gravitates down the slopes and slowly accumulates in the abysses beyond.

Where the coast is low it may be very old. In the absence of earth movements up or down, such coasts may remain static through long ages. The exception of course is the delta, where the shore is not static but actually growing outwards. The dissection of the land by deep bays and gulfs may result from subsidence of the land or from eustatic movements of the sea. Port Jackson and other harbours have already been shown to be drowned river valleys; Port Phillip, St Vincent Gulf and Spencer Gulf have in addition been produced by subsidence. Drowned river valleys are however so common round the world's coasts that some general explanation of the way in which they were cut below present sea level is called for. The usual view is that this took place during the times of low sea level during Pleistocene glaciations. But it has been doubted recently whether these times were long enough for the enormous amount of erosion involved. In the late Tertiary, in the Pliocene Epoch, it seems possible that subsidence of a large ocean floor may have lowered the world-wide level of the sea by increasing the total capacity of the oceans for a long period. This, together with the later glacial low sea levels, could account for the cutting of the drowned valleys.

The problem of the continental shelf is much more complex. Why Australia should be surrounded by a shallow strip of sea bottom with great oceanic abysses beyond is a question difficult to answer. The continental shelf varies a great deal in width, from over 100 miles in central Queensland to from 20 to 30 miles in New South Wales. In New South Wales the sea bottom shelves gradually to depths of slightly more than 1,000 feet, then it descends by a significantly steeper slope for about 14,000 feet to the bottom of the Tasman Sea. This is a major physical feature about which there is very little information. Results from the new methods of exploring the sea floor have not yet appeared for the Australian continental shelf so this problem must be left. Suffice it to say here that work of this nature elsewhere has shown that there are several kinds of continental shelf in their structure and origins.

With these preliminary remarks in mind let us embark on the circumnavigation of Australia, pausing here and there for a brief examination of parts of the shoreline. This imaginary voyage starts at Brisbane, and the first stage is southwards along the coast of New South Wales.

Here, at the commencement of the journey, is a feature of unusual interest. Emerging from the estuary of the Brisbane River, the ship enters the broad shallow waters of Moreton Bay. Moreton Bay is not a bay in the true sense, but rather a long strait, running north and south for over 60 miles, and narrowing at its southern end, where it contracts to a tortuous channel. Separating it from the sea on its eastern side are two narrow islands, Moreton Island and Stradbroke Island, probably at one time continuous, but now separated by a small gap. To reach the open sea it is necessary either to negotiate the difficult and shallow southern passage or to make a long detour round the northern end of Moreton Island.

The presence of islands within the sea, as of lakes within the land, always has some special significance. In a way both are abnormal to their surroundings: they are due to some peculiar and temporary combination of circumstances, and under normal conditions are destined to disappear. These islands are no exception to the rule, nor is Fraser or Great Sandy Island, a little farther to the north.

At first sight there is nothing very distinctive about them. They rise some hundreds of feet above the sea; there are valleys between the hills, which are covered with vegetation, including many large trees. On Fraser Island there are even some patches of real rain forest. The highest point, on Moreton Island, is Mt Tempest, 919 feet above sea level.

Fraser Island is in shape and setting curiously like the other two. It is still larger, narrow and over 70 miles long, and very close to the mainland at its southern end, thus forming a large V with the coast. Partially enclosed are the waters of Hervey Bay, which, like Moreton Bay, is wide open to the north.

The peculiar feature of these islands is that they are composed entirely of sand. In the whole of their extent there are only a few small areas of bedrock. They are in fact glorified sandhills or dunes, and Mt Tempest is probably at the moment the highest sand dune in the world. At Double Island Point on the nearby mainland there are dunes which form part of the same chain of dunes. Wood and charcoal from the uppermost part of this dune sequence have yielded various dates going back to material older than 45,000 years, about the limit to which radiocarbon dating can go. There are other dunes much older than these dated ones and coastal dune formation may have gone on here from time to time throughout the Pleistocene.

On the seaward side numerous patches of white above the waterline mark where the waves are gouging their way in to undermine the tree-clad slopes above. Even within the last few years considerable areas at the southern end of Stradbroke Island have been worn away and are now but shallow sandbanks awash with the tides. This shows that conditions since the building of the islands have changed, and with their changing it is reasonable to assume that the process of destruction will continue until all are gone. Large parts are, however, being radically transformed nowadays at a rapid rate by the rutile miners.

Sand dunes are really complex formations. A dune technically is a hill composed of wind-blown sand. In desert regions the prevailing winds heap sand into hills which assume a variety of definite patterns. The variation in air currents causing these patterns is still somewhat obscure, and it is hard to understand just why dunes in one area appear as long parallel hills, while in another they are shaped like horse-shoes and are quite separate from each other. Apart from deserts, dunes are common on sea coasts, particularly where a coast is flat for some distance inland. Sand washed upon the shore becomes dry at low tide, and sea breezes blow it inland, where it accumulates as sand hills. Such dunes are being built in dozens of places along the seashore at the present time. Active dunes several hundred feet high are known today and the fossil dunes of Fraser and Moreton Islands are not grossly different in size. These must have been formed in the same way. Some geologists have claimed the climate was different with drier conditions or stronger winds, but this does not seem necessary for their formation. However there is no doubt that sea level has gone down and up many times and these changes may be involved in the several phases of dune formation known to have occurred. Dunes may have started forming east of the present coastline and have been driven westward with rising sea level to end up where they are today

Between Hervey Bay and Moreton Bay is another large series of old sand dunes closely resembling those of Fraser and the other islands, but still forming part of the mainland. These lie behind and parallel to Teewah Beach north of Noosa Head, and their hinterland is a large coastal plain partially covered with lagoons. Like the others they are old and tree-covered; and they justify the term "fossil dune", which has often been used.

Leaving the sand dunes of Moreton Island and sailing south on our imaginary voyage, the coasts of New South Wales, north-eastern Tasmania and eastern Victoria, though differing in detail, will be found to have certain features in common. The whole of this coast has few inlets and no large gulfs, and is broken only by the shallow estuaries of the coastal rivers. There are also no large islands and only a few small rocky islets quite close to the shore. The continental shelf is quite narrow, varying from 20 to 30 miles, and sloping gradually to about 300 fathoms before it steepens to the tremendous escarpment at the edge of the abyss of the Tasman Sea nearly 3,000 fathoms below.

The Tasman Sea has been in existence for a very long time, and also the present coastline has been in nearly the same position at least since the Triassic times. The last great land movement took place much later in the Tertiary, when the eastern highlands were uplifted into a tableland, and this may have been accompanied in the south, as it certainly was in northern Queensland, by subsidence on the seaward side. Since then the coast has been static, its details modified only slightly by sea erosion and eustatic movements of the sea throughout and after the Glacial Age. In the chapter on Port Jackson it has been shown how the last rising of the sea inundated a river valley and turned it into a spacious harbour. At the same time other river valleys became bays, such as Broken Bay and Port Hacking, while low-lying parts of the coast were also inundated to become shallow bays, such as Port Stephens, Lake Macquarie, Botany Bay, Jervis Bay and Twofold Bay.

In the last few thousand years many small islands have been reunited to the mainland and parts of the platform eroded by the sea converted into coastal plains, swamps and lagoons (Figure 24d). This has been attributed to a slight fall in sea level in the last 5,000 years but the latest investigations have failed to show that this did certainly happen. Wave action constructing sand barriers is capable of bringing these things about without sea level changing to help it

Most of the present detail of coastal topography has resulted from erosion by the sea on different rock formations. There is a great diversity of rocks in south-eastern Australia, and each produces its own type of coastal scenery.

Near the mouth of the Clarence River, also from Newcastle south to Jervis Bay, and on parts of the Tasmanian coast, are horizontally bedded sandstones and shales of the Permian and Triassic periods. The massive sandstones of the Clarence River and Sydney are nearly homogeneous in structure, and when undercut by the waves form high vertical cliffs. The cliffs generally descend directly into the sea, though occasionally a harder layer may allow a narrow ledge to develop at their base. More often there is a heap of talus, masses of sandstone fallen from above, piled one upon the other, temporarily acting as a buffer to the waves, and protecting the cliffs from further erosion until such time as they too disintegrate (Figure 24, *a*).

Below the Sydney sandstones is a series of shales, sandstones and conglomerates known as the Narrabeen beds, and below these again are more sandstones and shales of the Permian coal measures. Both formations outcrop on the coast to the north and south of Sydney, and the presence of much soft shale interstratified with the sandstone considerably alters the scenery. Cliffs are not always so high, contours are softened, and there is often a wide flat platform uncovered at low tide. Where there is a hard sandstone layer just below high tide level the sea may cut the overlying softer rock back for a considerable distance. The platform may then be very wide as at Long Reef just north of Sydney, and as here the strata dip slightly towards the land, the seaward side is high enough to be uncovered at high tide, leaving a channel between it and the shore. Many such platforms are to be found along the coast between Sydney and Newcastle as well as in the Illawarra district to the south (Figure 24, *c*).

In the extreme north of New South Wales much of the coast is formed of very hard quartzites of doubtful age, which have been intensely folded in a north-south direction. These make a very rugged shore, its appearance differing according to whether the strata dip towards the sea or the land. Cliffs are low but very rough, there is no rocky platform bordering the sea, and the water is studded with reefs separated by deep and dangerous channels (Figure 24, *e*).

Between the headlands in many places are wide sandy beaches backed. by lines of sand dunes, behind which are swamps and shallow lagoons. Many of the beaches are of pure white sand, others when exposed at low tide are wide beds of rounded water-worn pebbles, débris from the hardest parts of the worn coast. At and above high tide are the famous black sands, natural concentrates of heavy metallic particles, formerly widely dispersed through the local rocks. These sands have been profitably worked for such rare minerals as zircon, rutile and monazite, as well as the metals gold and platinum.

Most of the swamps and lagoons, here as elsewhere, occupy areas where there had been once softer rocks, since worn away by the sea. When the sea had drained from the rocky platform thus formed, beaches built up on its outer edge, and the sand at low tide was blown inland to form dunes. Behind the dunes the drainage was dammed and swamps formed, occasionally fresh, but more often brackish and with some outlet to the sea (Figure 24, *d*).

Midway on the New South Wales coast, where rocks of Carboniferous age form the shore, the topography is somewhat similar. These rocks are mainly folded sandstone and quartzite and are not quite so hard as those farther north. Here and there are great intrusions of porphyry which had originally forced their way upwards through the sandstone and cooled into vertical

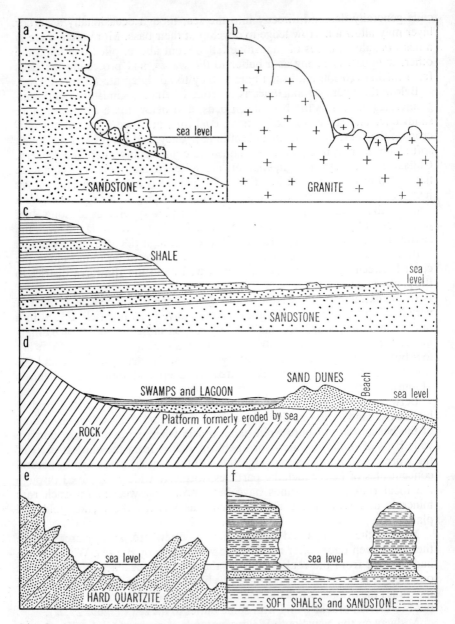

24. Sections of types of coastal scenery: (a) horizontal sandstone, Sydney; (b) granite; (c) sandstone and shale with slight shoreward dip, Narrabeen, New South Wales; (d) formation of coastal lagoons; (e) steeply dipping quartzite, Woolgoolga, New South Wales; (f) weak marine strata, Port Campbell, Victoria.

Sea cliffs and stacks in sandstone, Port Campbell, Victoria. *Victorian Railways.*

Granite coast near Albany, Western Australia. The inlet is a small drowned valley. *Western Australian Tourist Development Authority.*

Living coral, Hayman Isalnd, Queensland. *Queensland Railways*.

Coral reef and sand cay, Heron Island, Queensland. Artificial channel cut in reef on right.

pipes. Their extreme hardness has enabled them to resist erosion, and they now stand above the surrounding country, often as monoliths hundreds of feet high. Some which had previously been inland were converted into islands during the rising of the sea, and of these some were reunited with the mainland in late Recent times. The high rocks and islands at the entrance to Port Stephens are of this type, as are the lofty promontories of Cape Byron and Smoky Cape.

Granite often produces another type of coastline. There are few granite outcrops on the northern New South Wales coast, but it is more abundant on the south coast, on the east coast of Victoria and on the east of Tasmania. Granite varies a great deal in its resistance to erosion and decomposition, but generally it tends to weather into large smooth rounded masses, separated by deep clefts where joints have formed lines of weakness. Granite headlands are often high and bold, but they are usually rounded and are rarely flanked by vertical cliffs, for surface weathering rounds off the sharp edges and keeps time with the erosion of the sea from below (Figure 24, *b*). The few small islands off the south coast are chiefly of granite. Montagu Island consists of a mass of low rounded hummocks, and from Gabo Island comes the red granite formerly used in many buildings in the chief cities. Maria Island, off the Tasmanian coast, on the other hand, is largely composed of sandstone full of extinct fossil shells of the Permian Period. These islands, like those off Port Stephens, are residuals of harder rock, once part of the mainland, but separated by the final rising of the sea.

All down the east coast of Australia from southern Queensland to Gabo Island the unbroken coastline is backed by high land coming very close to the sea. This is in keeping with the rising of the eastern tablelands during the latter part of the Tertiary Period. The first major break is Bass Strait, originally a sunken area, where the main changes have since been due to changes in the level of the sea. Tasmania again is a land that has risen, its elevation coinciding with that of the Australian coast to the north. Its east coast is nevertheless more broken than that of New South Wales, a character becoming more emphasized as the south-eastern tip is reached.

Here is coastal scenery unlike that of any part of Australia. Storm Bay is a great gulf biting deep into the land; Tasman Peninsula is a long irregular deeply indented extension of the land; the winding D'Entrecasteaux Channel separates Bruny and other islands from the mainland. Another interesting fact is that the Australasian Antarctic Expedition of 1911-14, when sounding south of Tasmania, discovered that a submarine ridge extends far into the Southern Ocean. The average depth of the Southern Ocean is over 2,000 fathoms, but this ridge over 300 miles south of land rises to within 600 fathoms of the surface.

These are all the attributes of a sunken land. It is logical to conclude that the upward movement of the Australian eastern tableland was an undulatory one, like a series of waves of unequal height. The Kosciusko Plateau represents the crest of the highest wave, and there were troughs in central Queensland and in Torres Strait. Northern Tasmania was on the crest of another

high wave; its southern portion was not only on the sloping side but was
actually sinking into a great trough below the Southern Ocean. The submarine ridge far to the south may well represent another crest in this earth
wave which elevated the sea bottom nearly 9,000 feet but left it still over
3,000 feet below the level of the sea.

Coastal scenery in all sunken areas is generally spectacular. The sea strikes
fiercely at the remnants of high land remaining as islands and peninsulas
and carves them into striking headlands and rugged coves. In a region rich
in such features few are more impressive than Cape Raoul. It is on the
extreme south of Tasman Peninsula, and ships entering Storm Bay on their
course to Hobart pass close to its rugged face. The main mass rises in a sheer
cliff of black dolerite over 600 feet above the sea, a mass composed of closely
packed vertical hexagonal columns, continuous from top to bottom. On its
eastern side, where erosion has been greater, the cliff is lower and more
broken, the columns become less compact and stand above the skyline in
groups, while on the extreme tip the columns are completely separated and
stand individually in slender vertical stacks. Other lofty cliffs of columnar
dolerite occur on the coast farther to the south, on Bruny Island and on
South-west Cape at the southern tip of Tasmania.

The occurrence of these dolerites at sea level is further evidence of the
subsidence of eastern and southern Tasmania. In the previous chapter it was
described how sheets of dolerite form not only much of the surface of the
northern tableland but also the highest of the individual mountains. They
may have been originally intruded at different levels, but it is noticeable that
nowhere in the island are they so low as in the south, and it would seem that
the whole of this part has sunk. The dolerites may well continue and form
the submerged sea bed much farther to the south.

The coast at Eaglehawk Neck at the base of the Tasman Peninsula is
rugged and picturesque. The rocks are sandstones of Permian age and the
nature of the shore is similar to that of the Illawarra district and at Ulladulla
as described already. The joints are very regular, vertical and at right angles.
In the famous Tessellated Pavement the jointing is so even that the surface
of the eroded marine platform has the appearance of artificial masonry.
Elsewhere the sea has worn its way along master joints deep into the high
cliffs, and then laterally along other joints at right angles to the first. At
Tasman's Arch the head of what had at one time been a cavern has collapsed,
leaving a lofty arch bridging an awe-inspiring chasm. At the Devil's Coachhouse no arch remains; there is only a profound cleft, with a seething
cauldron of broken water below.

The largest inlets of the western coast of Tasmania, Macquarie Harbour
and Port Davey, have already been mentioned; they are very fine drowned
valleys. In spite of its loneliness, or perhaps because of it, Port Davey is a
gem of coastal scenery. Though only 50 miles from Hobart, it is still far from
the beaten track, virtually inaccessible by land and reached with difficulty
over a storm-tossed sea. Whalers, fishermen and timbergetters have made it
their headquarters for brief intervals and have in turn departed, and its shores

are uninhabited and in a virgin state. From the sea Port Davey opens into a fair-sized bay. Just within the entrance is the Breaksea Group, a chain of rocky islets populated with seals and penguins. These islets enable Port Davey to provide a sheltered anchorage with deep water right to the shore. From the main bay one arm penetrates the hills for about four miles to the north; another arm, Bathurst Harbour, runs for about the same distance to the east. Everywhere the mountains come down to the shore. Steep though the hinterland is, it is a region of rounded outlines, of black rocks and soft green foliage, of delicate half tones and horizons dimmed by mist.

From western Tasmania let us return on our voyage once again into Bass Strait. The general story of its many vicissitudes has already been told; there remains some consideration of the present scenery of its coasts. In the first great subsidence early in the Tertiary Period the strait was much wider than it is now, and the sea spread deeply into southern Victoria. To the south it did not encroach far beyond the present coastline, and the only place where Tertiary marine rocks occur is at Table Cape. This was the first Bass Strait, much wider and deeper than it is now.

The sea was deep enough to submerge a chain of hills formerly existing on its eastern side, hills of older rocks which were later to emerge as the Furneaux and Kent groups of islands. Another chain of similar hills existed on the west, and of these King Island and the neighbouring reefs are restored fragments of old land. The sea was also deep enough to allow the accumulation of a great thickness of marine rocks. How thick the rocks are where they are still buried beneath the sea is not known; but in the general elevation which followed, many were lifted above the surface and are now exposed in sections on many parts of the Victorian coast. Bores have there shown them to be over 1,000 feet thick, and they are probably much thicker below the middle of the Strait. At this time, when Tasmania was again for a while reunited with the mainland, the isthmus was broad with high east and west shores; and a low-lying area, possibly a large lake, occupied the centre.

The elevation of the sea bed, though considerable, was not great enough to bring it much above sea level even during the greatest recession of the sea in the Glacial Age. The fluctuations of sea level converted the strait to an isthmus several times before the final flooding gave it its present form. The present coastal scenery conforms generally to what might be expected in a sunken area. There are certainly no great indentations, for the original sinking as well as the elevation which followed, particularly along the northern Tasmanian coast, seems to have been along a fairly even line. This was probably a great fault. It is the archipelagos that reveal most of the subsidence: they are rugged islands of very old rocks, now deeply cut into by the sea.

Some islands, such as Flinders Island, Cape Barren Island and King Island to the west, are large and high, but others are merely the extreme summits of former hills rising in isolated peaks above the sea. Many of the old hills do not even reach the surface but lie awash as dangerous reefs. Others are still joined to the mainland as high promontories.

On the Victorian coast Wilson's Promontory projects as a great bastion of granite well into the Strait, its rugged cliffs still defying the fury of the stormy waters breaking at their feet. This is the northern part of the higher ridge which once continued right to Tasmania. Another high point on the same ridge is just off the end of Wilson's Promontory. This is Rodondo Island, a monolith of red granite, rising almost sheer to 1,150 feet above the sea. Well out in the Strait and about half-way to the little Kent Group is another similar monolith of granite, Hogan Island. On the larger islands, such as Flinders and King, there is proof that these were completely submerged in the first great subsidence, for lying above the older rocks are the remains of the marine sediments deposited in the Tertiary Sea.

East of Wilson's Promontory are the Gippsland Lakes behind the Ninety Mile Beach. Along the Latrobe valley the crust has continued to subside from the beginnings of the formation of Bass Strait before the start of the Tertiary. However the large rivers from the Eastern Highlands of Victoria have been able to keep pace usually and kept it filled with sediment; there are wide alluvial plains today. On the seaward side the embayment is cut off from the sea by long curving sand barriers. On the outermost is the Ninety Mile Beach. These have been formed by ocean waves from the south-east with sand dunes piled on top by the wind. Into the shallow waters of the lakes enclosed behind, rivers like the Mitchell, Tambo and Latrobe have built deltas, projecting in some cases like fingers into the water. At the back of the lakes also there are some low hills of friable sandstone; these are old solidified dunes like those behind Noosa Head in Queensland mentioned earlier.

From Cape Otway in Victoria and far to the west there is a great change in the coastal scenery. The sediments deposited in the Tertiary Sea were here elevated, so that their uppermost beds are now as much as 500 feet above sea level. They form not only most of the coast but underlie the western plain of Victoria, south-western New South Wales, parts of South Australia and the hinterland of the Great Australian Bight.

The centre of Otway Peninsula consists of Jurassic rocks and is over 1,000 feet high. It was probably an island in the Tertiary Sea.

The Tertiary rocks form the coast on both sides of the peninsula at the present day. Most of them are rather soft, consisting of loosely consolidated beds of clay, sand and limestone. The sea is cutting into them, and as the coast rapidly recedes there is left a wide shallow platform at the vertical limit of the power of the waves. The shore is bordered by vertical cliffs, sometimes of considerable height, descending straight into the sea. Here the fallen material rapidly disintegrates, and there is no pile of large fallen boulders as at the foot of the cliffs of harder sandstone in New South Wales. So rapid has been the advance of the sea that slightly harder masses of rock have been left isolated well away from the shore. They may be seen in all stages of disintegration, some in large blocks and moderately intact, others as needle-like spires, others as rounded hummocks awash with the waves. Near Port Campbell on the west side of the Otway Peninsula a long straggling and picturesque line of these residuals is known as "The Twelve Apostles" (Figure 24, *f*).

The sea cliffs extend along most of the western Victorian coast, but near and across the South Australian border they become lower and are replaced by long sandy beaches, backed by a broad zone of parallel sand dunes. The dunes include much calcareous sand and shell; the older dune ridges have been converted by slow seepage into impure limestones. Behind them again is a zone of low-lying ground with many swamps and lagoons, covered usually with a dense scrub of tea-tree. The delta of the Murray River lies inland of the largest coastal lagoon, Lake Alexandrina, but it is a modest feature. So little of the flow of the Murray ever reached the sea across its long dry plains that it was not a very effective delta builder although it drains the sixth largest drainage basin in the world. The tortuous channel through the sand dune ridges on the seaward side of Lake Alexandrina is the way out to the sea for the occasional floods of the Murray big enough to get this far. At low sea levels this winding course must have been that of a freshwater river; at times of high sea level it must have been an estuarine channel as indeed it was until very recently. But now a barrage has been built across it and it is just an arm of Lake Alexandrina which has been turned into a freshwater lake.

Westward the old rocks of the Mt Lofty Range impinge upon the sea, and the coastal topography again changes. Some of these rocks here and on Kangaroo Island are granites, granites which were old land surfaces far back in geological history. Until recently they had been buried beneath glacial deposits of the Permian Period, and now on the coast these are being worn away, leaving the granite hills much as they were after the Permian ice sheet had passed over them more than 230 million years ago.

St Vincent Gulf and Spencer Gulf both lie in rift valleys that were partially submerged by the last great rising of the sea. The strait separating Kangaroo Island from the mainland is probably a fault-guided valley running at right angles to the others. The core of Kangaroo Island is granite, similar to that of the mainland, and the sea has worn the granite on the coast into many picturesque formations. Huge rounded boulders of granite are in places piled one upon the other, similar in appearance to the "Giants' Marbles" of central Australia, but many of the boulders are eaten into and honyecombed with innumerable miniature caves and grottoes.

Farther still to the west the coast is similar to that of western Victoria. Along the whole of the Great Australian Bight there is remarkable uniformity, and the shore is bordered by a continuous line of high cliffs composed of limestone deposited in the Tertiary Sea. The face of the cliffs undulates in many wide curved bays and headlands, although on any but the largest-scale map it appears as an unbroken line. In only one part do the cliffs recede from the sea, and here in the middle of the Bight there is for some distance a coastal plain up to 20 miles wide. The cliffs are from 100 to 150 feet high, and from some distance at sea they form an even skyline, giving rise to the erroneous term Hampton Range, a name they still bear. Above and behind the cliffs is the Nullarbor Plain, described in another chapter. These cliffs are simply emerged sea cliffs in continuation of the present sea cliffs to the east and west; sea shells occur widely in beds at the foot of the cliffline behind the coastal plain. The coastal plain was produced by marine erosion cutting back into the limestones to form the inland cliffs and is not due to downward faulting as has been suggested.

From Israelite Bay to Albany the coast consists of ancient rocks, granites and gneisses once part of the shadowy land of Yilgarnia. It was probably in Tertiary times that the southern portions of Yilgarnia subsided, until part of it was below the sea for the first time since the dawn of things. Higher parts, such as the coastal hills at Albany, had been islands in the Tertiary Sea, for the flat country on their landward side is underlain by marine rocks of Miocene age. As the sea level fell and rose in Pleistocene and Recent times the granite hills had a chequered career, sometimes forming lofty promontories, sometimes being surrounded by a coastal plain, sometimes being islands distant from the shore. The many small islands and granite islets of the Recherche Archipelago are all part of the same formation, isolated by the last general rising of the sea. There is here invariably the typical coastal scenery of granite and allied rocks, high steep-sided hills with rounded summits, sometimes consisting of one large monolith, but more often divided by deep fissures and joint cracks.

Eastern and Western Australia present many structural contrasts, but none is more apparent than the scenery of the coast. The most striking feature of the Western Australian coast is its uniformity. From Bremer Bay, which faces the Southern Ocean for more than 150 miles to the extreme southwestern point, thence northwards to beyond Shark Bay, for a total distance of more than 1,000 miles, there is practically no change. A castaway upon any portion of this far-reaching shore would find no clue in its physical features to show where he was stranded.

This uniformity is due to several causes, one the evenness of the coastal plain, another its gradual slope downwards to the edge of the continental shelf, 100 miles from the land and 100 fathoms beneath the surface of the sea. Throughout the whole of this distance there has accumulated in the course of time a peculiar formation known as the coastal limestone. This belt of limestone is narrow, sometimes only a few yards in width, but sometimes it extends for many miles inland. It is generally low, but it may form high cliffs; at one place north of Geraldton a line of cliffs rises to nearly 700 feet above sea level. It has also been proved by bores to run more than 200 feet below sea level, and it undoubtedly extends well out to the continental shelf.

Some of the limestone is Recent in age, but much of it formed during the Pleistocene Epoch. In the last intense phase of the Great Ice Age, when the sea was over 400 feet lower, the west coast of Australia extended some miles farther into the Indian Ocean. During the rising of the sea, and assisted by the growing aridity of the climate, the wind began to build large sand dunes along the coast. Some of the earlier dunes were submerged by the sea, which finally

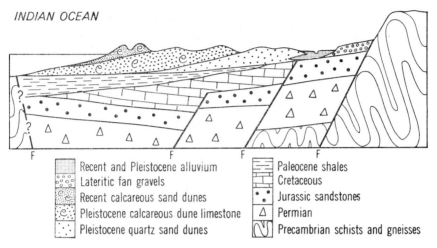

DARLING RANGE

INDIAN OCEAN

Recent and Pleistocene alluvium
Lateritic fan gravels
Recent calcareous sand dunes
Pleistocene calcareous dune limestone
Pleistocene quartz sand dunes
Paleocene shales
Cretaceous
Jurassic sandstones
Permian
Precambrian schists and gneisses

25. Sections across the coastal plain of south-western Western Australia.

rose to about its present level. Layers of sea shells were deposited above the older dunes; then new dunes formed above the old ones. Apart from the included sea shells, much of the wind-blown sand consisted of small particles of shell ground to pieces by the surf. In the course of time rain water percolating through the dunes dissolved and redeposited the carbonate of lime forming the shells, and consolidated the whole into a compact but rather impure limestone.

This is the limestone which everywhere forms the coast. The sea has already eroded drastically the outermost dune ridge, and left portions such as Garden Island and Rottnest Island as much as ten miles from the shore.

A limestone coast has its own peculiarities. This one is for the most part low, picturesque and exceedingly rugged. From below it is eaten into by the sea, from above it is corroded by the rain. The sea carves it into caves and grottoes. Harder parts are so undercut that they remain perched on narrow pedestals and have earned the appropriate local name of "mushroom rocks", or they may be worn through to form natural arches and bridges. From above the rain trickles down the sides of the rocks and corrodes fluted channels, or seeps through small holes and enlarges them to vertical pipes. There are sharp pinnacles where pipes of harder limestone have remained after the surrounding material has been dissolved away. These pipes formed beneath a soil cover; water percolated downwards, dissolving the calcareous sand and reprecipitating lime around the developing cylinder into which soil and roots came down from the surface. Their formation is part of the general process consolidating the dune sand into limestone. Later the retreat of the coast under wave attack brought some of these pipes into the domain of sea spray as well as rain. The soil and loose materials are removed to expose hollow pipes. The surface of the rock is always rough, sometimes pitted, sometimes corroded to a perfect honeycomb. Everywhere are sharp knife edges, difficult to traverse and fatal to the toughest of boots.

The ruggedness of the rocks is not confined to the land but is continued on the adjacent sea bottom. Owing to the gradual slope of the continental shelf the sea is shallow for some distance from the land, its maximum depth dependent upon how far the waves have worn or corroded the underlying limestone. It is a dangerous and treacherous zone for navigation, studded with wave-worn rocks and jagged submerged reefs. There are few inlets except the estuaries of the rivers, until Shark Bay is reached near the northern limit of the coastal limestone. The long low peninsulas enclosing this shallow bay are of limestone, and are already breaking up into islands before the corrosion of the sea.

Continuing our imaginary voyage round the north-west of Australia, and passing the Eighty Mile Beach, where the central desert impinges upon the coast, we come to a region different from any other part of Australia. It is also remote and off the beaten track of shipping, visited only by iron ore ships, cattle boats, pearlers or more rarely by Admiralty survey ships and scientific expeditions. A volume could and should be written about its scenic wonders.

As revealed by the map the coast is rugged. Between Broome and Wyndham it is indented by deep gulfs. It is flanked by the innumerable small islands and reefs of the Bonaparte Archipelago. The neighbouring sea is shallow, and small islands and reefs break the surface right across the Timor Sea to the island of Timor itself. It is a sea difficult and dangerous to navigate, and the danger is increased by the tremendous rise and fall of the tides. This varies, though it is always considerable. At Walcott Inlet it is only about 16 feet, at Wyndham it is 21 feet, at Prince Frederick Harbour 28 feet, and at Derby no less than 36 feet.

Of the many large inlets, the westernmost, King Sound, is the largest, an almost rectangular body of water, 65 miles long and 35 miles wide, separated from the sea by a fringe of small rocky islands and reefs. Among the islands on the eastern side is one now becoming prominent in Australian economics, an island composed of almost solid iron ore. This is the famous Yampi deposit, destined to contribute to our supplies of iron and steel for many years to come. Many other large gulfs indent the land to the east, none quite so large as King Sound. The largest are Collier Bay, Brunswick Bay, York Sound, Admiralty Gulf, Vansittart Bay, and finally, after a long stretch of coast, Cambridge Gulf.

Many of the harbours are deep, spacious and nearly land locked. It has been reported by Easton that many are flanked by precipitous cliffs, apparently of sandstone, rising hundreds of feet sheer from the water, making the shores inaccessible and the harbours useless. The scenery is very fine. In Prince Frederick Harbour the bluffs are high, of horizontally bedded red sandstone. Rothsay Harbour at the mouth of the Prince Regent River is somewhat similar. Farther inland, though the river valleys are moderately

Glory Arch, Yarrangobilly Caves, New South Wales. Part of the cave roof has collapsed behind the arch. *New South Wales Department of Tourism.*

Lake Cave, Margaret River, Western Australia. Stalactites, stalagmites, and pillars (one broken).

wide and flat, they are still flanked by high cliffs. The land above the cliffs is fairly level, rising gradually towards the interior.

The broken coastline of the Kimberleys suggests recent change. The simplest explanation of the harbours is that since the elevation the tableland has been deeply dissected by the numerous rivers, and that during the rising of the sea after the Great Ice Age the mouths of the valleys were submerged by the sea. The rising of the sea may well have been preceded by some subsidence of the coastal belt, and the numerous islands would then be the summits of hills formerly rising from a coastal plain.

J. T. Jutson, a very able geologist, says that the harbours are all drowned river valleys. The great size and rectangular shape of King Sound suggest that it may be a rift valley. However no faulting along its western side has been proven and that on its eastern appears to very old. Therefore King Sound seems to be simply a drowned feature also; however, what was drowned was not a simple valley but a broad plain eroded in weak rocks, one side of which was defined by an ancient fault. The land to the west of the Sound consists of spinifex-covered plain extending to Broome and far beyond, relieved only by mesas of sandstone deposited in lakes and seas in Jurassic times. This region would seem to have been stable for a long time and always at a low altitude, hence the coastline is very little broken and is marked by such features as the Eighty Miles Beach.

It is quite a long sea voyage from the Kimberleys to the most northerly point of Australia at Cape York, and the most remarkable feature of that journey is the one major indentation of the whole Australian coast, the Gulf of Carpentaria.

This great, almost rectangular sheet of water, some 400 miles across and penetrating the land for the same distance, is virtually a sea bounded on three sides by almost straight coastlines. Its topography is interesting and suggestive. On the eastern side lies Cape York Peninsula, a narrow northerly continuation of the eastern mountainous backbone. On the western side is Arnhem Land, not so rough and mountainous, but nevertheless a solid rocky bastion composed of very ancient rocks. To the south the sea shallows to a low marshy shoreland, the confluent deltas of a number of rivers. For mile upon mile are low mangrove swamps, infested by the lurking crocodile and myriads of greedy mosquitoes. The rivers enter the sea after winding sluggishly through a labyrinth of tortuous tidal channels, which may sometimes be traversed for 40 miles before dry land is reached.

Beyond the mangrove belt the hinterland is flat, and grassy plains with a few scattered trees stretch as far as the eye can see. The plains continue on seemingly for ever, their monotony relieved only by patches of the rock-like "white ant" hills, nests of the destructive termites, five or six feet high, and each species with its own distinctively shaped hill. It is possible by travelling south to traverse Australia without meeting hills of any magnitude. The Gulf of Carpentaria is thus a gateway to the heart of the continent, the northern termination of a great belt of low country practically dividing Australia into two halves. It is obvious that there must be some geographical link between

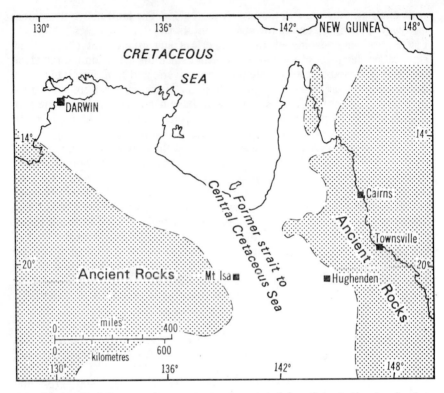

26. The Gulf of Carpentaria, gateway to the great plains of Australia, showing how the sea overflowed into Australia in the Cretaceous Period.

the Gulf and the low country to its south. The solution, as with so many topographical problems, lies in the past.

Fortunately the story is here easy to read. There was a time when the Gulf was a real gateway, a marine gateway, when it was an actual strait leading to a great central sea. Through three whole geological periods the great belt of country lying between mountains in the east and high rocky land in the centre of Australia was slowly and progressively subsiding. The subsidence was intermittent, at its deepest in the centre, at times suspended or even relieved by temporary intervals of slight elevation. In the initial stages large lakes formed, amongst the largest bodies of fresh water that ever existed in the history of the world. At this time in the Jurassic Period the Gulf country was just high enough to form a barrier between the lakes and the sea. Later, in the Cretaceous Period, further subsidence allowed the sea to pour in and cover much of central Queensland and western New South Wales and also parts of South Australia. Indeed at its greatest extension in early Cretaceous time the sea connected with the present location of the Nullarbor Plain, thus dividing the Eastern Highlands from the western parts of the continent.

South of the Gulf marine Cretaceous rocks underlie the whole of the plain, and fossil sea shells as well as the bones of enormous extinct swimming reptiles have been found in many places all the way to South Australia. These were all buried in the sand and mud which gradually shallowed and then filled the Cretaceous sea. Finally the land just south of the Gulf again rose slightly and the inland sea became isolated. It may have dried up completely for a while, but it is certain that it was again and for a long time converted into other vast freshwater lakes.

The southern shore of the Gulf of Carpentaria probably has much the same appearance as it had at the end of the Cretaceous Period about 65 million years ago. It is being built slowly outwards at the present time by the accretion of mud brought down by the rivers. The living things have changed, of course, for there have been vast evolutionary developments in the plants which once clothed the plains and the animals which roamed in their shade. However for short periods in the Pleistocene Epoch the geography was completely different. During the glacial low sea levels the coastline shifted tremendously over the shallow sea floor. Indeed the whole of the Gulf and nearly all of the Arafura Sea became land. Not only was New Guinea linked to Australia but land reached so close to Timor that only a narrow, but deep, strait remained between. It is possible that there was much migration of plants and animals in this direction as well as across the dry Torres Strait.

The Great Barrier Reef

Continuing our imaginary sea voyage and passing through Torres Strait to the east we enter the domain dominated by that very humble organism, the coral. No book on Australian scenery would be complete without some account of the Great Barrier Reef, yet this chapter is written with the utmost diffidence. Since James Cook in 1770 entered the dangerous waters within the reef it has been known as one of the scenic wonders of the world. Visitors have been attracted by its beauties and scientists by the wealth and variety of its life, to say nothing of the many problems of its age and structure. Few Australian subjects have been more written about, and a bibliography of works, both scientific and popular, would itself fill a volume. Hence my diffidence, for in one short chapter it is hopeless to try to emulate such literary classics as E. J. Banfield's *Confessions of a Beachcomber* and *My Tropic Isle,* or add to the information given by such writers as Saville Kent, Dr C. M. Yonge, T. C. Roughley, Professor W. Dakin, and many others.

I shall therefore adhere strictly to the theme of this book, which is to trace Australian scenery to its causes. Let us then epitomize the factors that have led to the building of the Reef.

The primary cause is the coral itself. Corals, though lowly animals of simple organization, are of ancient lineage and have played a tremendous part in geological history. Their remains have been found as fossils in rocks of all geological periods as far back as the Ordovician. They built large reefs as far back as the Silurian Period, something like 400 million years ago. These ancient reefs are still visible in many parts of eastern Australia. At Chillagoe in Queensland, at Tamworth, Molong and Yass in New South Wales, at Buchan in Victoria, even on the Gordon River in Tasmania, as well as in many other localities, it is possible to walk for miles in limestone country where every stone picked up will show the structure of an ancient piece of coral. These corals were true reef builders and may be considered the ancestors of those living today, though they belonged to genera, families and whole orders which have long since been extinct.

To study the structure of the coral animal it is easier for most visitors to the sea shore to examine its very near relative the sea anemone. Sea anemones can be found at low tide almost anywhere on the sides and bottoms of rock pools or beneath stones. Out of the water they appear as shapeless lumps of jelly, but under water they expand to cylindrical tubes surmounted by a crown of short tentacles, often beautifully coloured. Their anatomy is comparatively simple. The tentacles surround the mouth, which leads to the internal gastric cavity. Vertical partitions line the sides of this cavity, extending nearly to the centre. The tentacles are armed with stinging cells, and, as they wave gently, they create a current of water which brings tiny organisms within their reach. The food is passed through the gullet and digested in the central cavity, while undigested particles are ejected through the same passage. Sea anemones are entirely soft-bodied, but corals, with a similar organism, secrete a hard skeleton of carbonate of lime to support the outer wall and also the internal partitions. It is this hard skeleton that remains after the death of the animal and forms the main bulk of a coral reef.

Corals have not only a long range in time, but they are at the present day widely distributed throughout all the seas of the world, from the equator to the polar regions, and from low tide to the depths of the ocean. Not all are reef builders.

Some corals are small and cup-shaped; these are called simple corals, and the cup is the skeleton of a single polyp living a solitary existence. Other corals are compound, a curious natural assembly of a number of individual organisms into a large compound colony. Such colonies are found in other lowly animals, in the sponges, in many jelly-fish, and in some rather higher animals such as the bryozoa or lace corals. Though a colony is composed of many individuals, all are connected by living tissue, so that nutriment absorbed by one part is shared by the remainder. In some compound animals there is even a division of labour, so that individuals may vary from each other and be adapted for special functions such as reproduction.

Corals capable of building reefs are chiefly compound and are more restricted in their distribution. They grow only in warm tropical seas, in places where the water is clear and free from sediment, and at depths rarely exceeding 30 fathoms. Many of the individual polyps are very small, sometimes no bigger than a pin's head, but the colonies may be very large, often exceeding 20 feet in diameter. There is a tremendous variety of them, far too great to describe here, and the reader is referred for a more detailed account to such books as Dr C. M. Yonge's *A Year on the Great Barrier Reef* or T. C. Roughley's *Wonders of the Great Barrier Reef*.

It is sufficient to say that many are quite massive, particularly those living on the outer edge of the reef and exposed to violent wave action. Other colonies are smaller and spherical, some are encrusting, while those living in sheltered situations are often delicately branched. Most dead and bleached coral is white, but when alive the polyps glow with innumerable colours, rivalling the most brilliant of land flowers and making the reef a veritable garden of the sea. Associated with the compound corals on the reef are many simple ones, some, like the mushroom corals, quite large. There are also other organisms whose remains contribute to the building of the reef. There are calcareous seaweeds, sponges, shells of all kinds, the tubes of marine worms, starfish, sea urchins, lace corals and many others.

Coral reefs cannot grow where the temperature of the seawater is lower than about 68 degrees F., and this confines them very nearly to the equatorial belt bounded by the tropics of Capricorn and Cancer. They are also mainly found on the eastern side of large land masses, for on the western side there is often a current bearing cold water from the polar regions, and the temperature is too low for their growth. Moreover in equatorial latitudes most ocean currents set westwards, sweeping coral larvae which are free-floating plankton, towards the eastern shores of land masses and impoverishing western shores at the same time.

Three distinct types of coral reef may be recognized, fringing reefs, barrier reefs and atolls. It is with the last two that the greatest problems have arisen.

Fringing reefs are comparatively easy to explain. Off the shore of any land where conditions are suitable, corals grow gradually upwards from depths not exceeding 30 fathoms to the limits of low tide. Above this they cannot exist, for they die when exposed to the air for a very short time. Thus a continuous bank of coral is formed adjacent to the shore and just below tide level, its width depending upon the slope of the sea bottom to the extreme limit of coral growth. The upper surface of the bank will be a mass of living coral and other marine creatures, the lower portion the consolidated remains of countless former generations.

Should the sea bed be stable, and subject to neither elevation nor subsidence, broken material thrown upon it by the waves may raise it above sea level. Most of the coral will then die, but some will still grow outwards. The action of the surf on the outer edge continually breaks portions away, and these are washed down the slope into deeper water. Many other organisms live here, and so in course of time sufficient material accumulates to shallow the water and allow a new growth of coral to commence. When shallow banks exist away from land coral reefs may form and rise to the surface over a considerable area, and these may be classed with the true fringing reefs even if there be no adjacent shore.

Barrier reefs and atolls are more complex in their origin. A barrier reef is a long line of reef or reefs parallel to but at some distance from the shore, from which it is separated by a continuous shallow channel. On the seaward side it generally slopes down extremely steeply to the great ocean depths. Atolls are rings of coral reefs often situated in the midst of the ocean remote from any land. In the centre of the ring is a shallow lagoon corresponding to the inner channel of a barrier reef, and there is on the seaward side a similar precipitous slope down to the ocean depths. Both barrier reefs and atolls may be raised slightly above sea level to form a continuous island, they may be surmounted here and there by small isolated islets, or they may be awash and except for the breaking of the surf invisible from the surface. The line of reef is rarely quite continuous, and there are often gaps to give access between the channel or lagoon and the open sea. In barrier reefs the gaps are sometimes but not always opposite the mouths of rivers, in atolls they are nearly always on the side leeward to the prevailing winds.

As coral reefs cannot live at depths exceeding 30 fathoms, their presence far from land, where they apparently rise from depths of thousands of fathoms, naturally early aroused the interest of zoologists, and few subjects have led to more theorizing and controversy. In a limited space it is impossible to do more than give the briefest summary of the different theories and the arguments for and against that have been advanced by their various proponents.

A theory advanced in the early nineteenth century that atolls had been formed on the rims of the craters of submarine volcanoes was soon discounted. Since then there have been three main theories, Darwin's theory of subsidence published in 1842, Sir John Murray's theory that the coral had grown on a base of sediment deposited on the top of submarine mountains, and Professor R. A. Daly's theory that the growth of coral kept pace with the rising of the sea after the Glacial Age. Where there are many theories about the origin of some natural phenomenon it has often been later found that each has contributed something to the ultimate truth. The view is now generally held that Darwin's theory satisfactorily explains the existence of most barrier reefs and atolls, but that the other theories are true in part and explain many divergences from the general plan.

Darwin's theory is based on the recognized fact that many parts of the earth's surface, including the beds of oceans, have slowly subsided in recent times. By this subsidence many volcanic islands in the centre of the ocean have become submerged, and the margins of larger land masses have also become depressed beneath the sea. Figure 27 shows how during this submergence a fringing reef would be first converted to a barrier reef and then to an atoll.

In Figure 27, (a), (b) and (c) represent three different stages of sea level while an island is slowly sinking. When sea level was at (a) a fringing reef formed in the manner already described. When the island had sunk so that sea level was at (b) the fringing reef had grown upwards, keeping pace with the subsidence, and now formed a barrier reef at some distance from the shore. At a still later stage, when the island was completely submerged, the

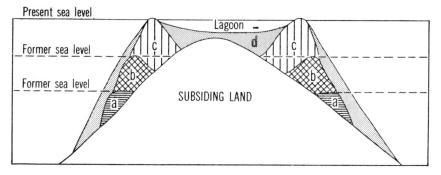

27. Section illustrating Darwin's theory of coral atolls: (a) fringing reef; (b) barrier reef; (c) atoll; (d) coral debris.

coral, still growing upwards, had reached sea level at (c), forming a complete ring or atoll with an interior lagoon. Debris broken by the waves has meanwhile rolled down the outer slopes, and on the inner side has with other marine growth partly filled the lagoon and kept it shallow. This is a very brief outline of Darwin's theory, though of course a great deal of detail and argument could be added.

Sir John Murray's theory claims that there has been no subsidence, but that when a submarine volcano has been built up within a certain distance from the surface, the remains of floating organisms, such as diatoms, sink and accumulate on the summit, raising it in time within the 30 fathoms necessary for the growth of corals. These then grow upwards to the surface. His explanation of how the atoll with its central lagoon is formed is an ingenious one based on the dissolving powers of sea water, but this side of his theory is definitely not borne out by other observed facts. On the other hand, the shallowing of the sea bottom by sedimentation in some localities has undoubtedly provided a suitable environment for coral growth.

Professor Daly's theory is really in keeping with Darwin's, except that it replaces subsidence by the rising of the actual sea level. These eustatic movements have undoubtedly played some part, but they are insufficient to explain the great depth of coral in the oceanic islands. The last great rising of the sea after the Great Ice Age had a maximum of about 400 feet; yet in a bore put down on Funa Futi Atoll, coral was proved to a depth of over 1,100 feet, and this would seem to indicate that the main essentials of Darwin's theory are correct.

Such then is a summary of the different theories of the origin of coral reefs. It remains to be seen just what part these theories played in building up that unique formation, the Great Barrier Reef itself.

The Great Barrier Reef, or rather reefs—for it is composed of many separate units—lies like a protective shield off the east coast of Queensland from the coast of New Guinea to the tropic of Capricorn, 250 miles from the border of New South Wales, in all a distance of 1,260 statute miles. It is a complex structure composed of many types of coral formation. The outer portion, the true barrier, is from six to over 80 miles from land, and stands like a narrow wall on the extreme edge of the continental shelf, its seaward face dipping almost vertically downwards to great depths. Within the barrier are the shallow coral-studded waters of the coastal channel. In the north the outer barrier for more than 600 miles is practically a continuous solid wall with very few gaps; but this wall becomes more and more broken, until in the extreme south it merges into scattered groups of small islands connected by a maze of wide shallow reefs.

Among the most southerly of these islands is the well-known and often visited Capricorn Group, just off Rockhampton. The Swain reefs lie a little to the north and are much more extensive if not so well known. They are a confused labyrinth of reef and island, intersected by tortuous coral-studded channels, and are in all about 50 miles wide and nearly 100 miles from the mainland. From here to Townsville the true barrier begins to appear, coming

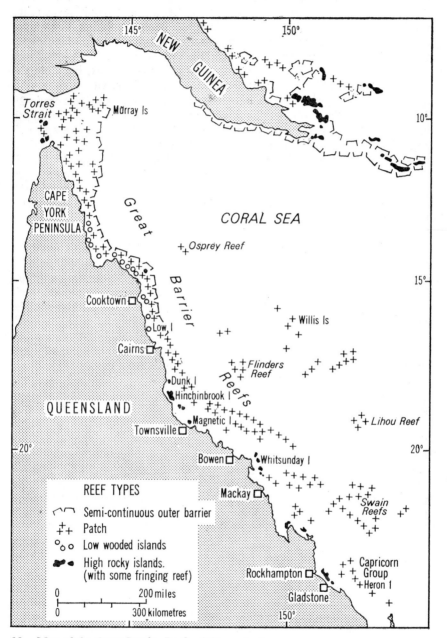

28. Map of the Great Barrier Reefs of Queensland.

nearer and nearer the shore. At Townsville it is 50 miles distant, at Cairns 20 miles, and at Cape Melville, 200 miles farther north, it is only seven miles away. Beyond Cape Melville the barrier is practically unbroken; it continues nearly due north, while the coast trends to the north-north-west. The channel thus again becomes wider and at Cape York is over 80 miles wide. From here to the New Guinea coast there is no land to the west, but only the recently submerged Torres Strait, itself a shallow reef-strewn sea.

The channel between the outer barrier and the shore rarely exceeds 20 fathoms in depth. Close to the shore, and particularly in the southern portions where the mainland mountains come down to the sea, are many high rocky islands. Most of them have narrow fringing coral reefs on part of their coasts, but few are completely surrounded by them. The high rocky islands are composed of similar rocks to the mainland, and are obviously detached portions which have been separated either by a general subsidence or by the rising of the sea. Many are of great scenic beauty, rising sheer from the water to heights above 1,000 feet, their sides and summits covered with dense forest and jungle.

Whitsunday Island near Bowen is the largest and is separated from the mainland by a long channel, the famous Whitsunday Passage, much used by coastal shipping. Whitsunday Island and the still loftier Hinchinbrook Island farther to the north are essentially part of the mainland, and their story has already been told in the chapter on the eastern highlands. The smaller Dunk Island has been immortalized as the home of E. J. Banfield, and through his writings is better known than any other island on the coast. Rocky islands are not so common in the extreme north, but Lizard Island, near Cooktown, is the unusual distance of 16 miles from the shore.

Between the zone of rocky islands and the outer barrier large areas of the sea bottom are covered with sand, but there are also innumerable reefs of living coral rising to variable distances below the surface. Upon the higher of these are the cays, low sandy islets rarely more than a few feet above high tide mark. Cays are similar to the islets found on the rims of atolls in mid-ocean. They begin with the accumulation of sand, broken coral and other débris thrown by the waves on the surface of a reef. At extreme low tides the action of the wind may help to pile the débris higher, until at last it it above reach of the highest tides. Seeds that drift ashore or are brought by birds take root, and their ultimate decay contributes to a growing layer of soil. Birds nest in the trees or upon the ground, and their guano adds to the bulk of the new island.

A curious feature of some of the reefs is that two distinct types of islets may exist side by side, a sand cay to windward and a mangrove swamp to leeward. Wherever the ubiquitous mangrove finds a footing it multiplies apace, and its tangled roots make a natural trap for minute floating particles of mud. Soon a dense impenetrable swamp develops, similar to those on the shore of the mainland a few miles away.

The mangrove swamp is a little world of its own, so dense that even the strong sea breezes cannot penetrate more than a few yards into its secluded depths. The air is still and heavy with the smell of mud and decaying vegetation. Here nevertheless is an abundance of life not found elsewhere on the reef, life especially adapted for an existence in the mud. Whelks crawl on the mud, oysters and other bivalves are attached to the roots, there are worms, crabs and fish and innumerable other living things both on and under the surface.

The headquarters of the Great Barrier Expedition of 1929, led by Dr C. M. Yonge, was on Low Island, a cay on a reef that also possessed a mangrove swamp. This reef, like others, had another feature, a narrow bed of boulders forming an almost continuous rampart round its outer edge. The boulders forming the bed are of water-worn fragments of coral cast up from time to time by storms. Grains of coral sand have acted as a cement and eventually the whole has been converted into a solid conglomerate. Within the circle thus made the wide flat of the reef surface is covered with water at ordinary low tides, though a few parts are uncovered in the lowest spring tides. The whole thus presents very nearly the appearance of one of the true atolls of the open ocean. There are innumerable cays similar to Low Island throughout the region, some like Heron Island without the accompanying mangrove swamp, and others with or without surrounding boulder beds.

Quite a distinct type of island is found in the extreme north, east of Torres Strait, but still within the outer barrier. The three Murray Islands and Darnley Island are all extinct volcanoes. Such islands have no relation to the mainland at all, but are similar to the oceanic islands which are the summits of volcanoes erupted from the ocean bed, sometimes many thousands of fathoms below. How long it is since the volcanoes became extinct is not known, but it must have been within comparatively recent times. On the island of Mer the original crater is still intact and shows little signs of weathering, and it is quite possible that its fires were once visible to human eyes. As with other islands within the barrier there is around Mer the usual fringing reef.

The outer barrier, the true barrier, is perhaps the most interesting; it is certainly the least accessible. Its approach is difficult even in the calmest weather, and portions only are visible during the lowest spring tides. Coming from the landward side the first sign of its presence is the roar of the surf and a long line of white upon the horizon, where the rollers of the Pacific rise up and shatter themselves upon the reef. On the inner margin the water is very shallow for some distance, and the bottom is covered with living coral. The surface of the barrier itself is nearly devoid of life, particularly where it is above the level of low spring tides. In such places it is smooth, hard and polished, and may run for miles like a broad concrete road.

As the tide rises the waves sweep right across, but the smooth surface offers no obstruction and erosion is probably very slow. The deep waters of the open ocean, unlike the coastal breakers which cut deeply into the hardest cliffs, carry no sand or other sediment to act as cutting tools. Such cracks and hollows as do develop are soon filled and repaired by living things. One of these is the hardy Lithothamnion, an encrusting calcareous seaweed, others are small corals that live in the occasional pools, and large sessile tubes, which are really the shells of a mollusc that does not follow the usual spiral habit.

If the surface of the barrier is dead, the steep seaward face a few feet below the water is alive with a host of sea creatures. This is a world in itself, and largely an unknown world, for only an occasional glimpse of what lives there can be obtained, and naturalists can only speculate as to the new and rare forms awaiting identification.

This then is the real barrier, remote from the shore, a stony rampart between the open ocean and the tropical lagoon. It is here that something is seen of the endless struggle between the growth of a reef and its destruction by natural forces. It might be imagined that a reef grows steadily upwards at a rate governed only by the rate of coral growth. Actually it is a race against time, and the question is whether the reef can build faster than it is destroyed.

The most apparent destroying agent is the sea itself. Waves are for ever tearing portions of the coral away and grinding them to powder. This goes on all the time, but during violent storms it is increased a hundredfold. The well-known nigger heads are large blocks of coral that have been detached and hurled on top of the reef by the waves; but they are a mere fraction of the material washed away and distributed on the neighbouring sea bottom. Within the reef itself are many organisms which corrode and destroy. Boring mussels, sponges, certain algae and other organisms not only perforate the dead coral, but so weaken it that it is often readily detached by the waves.

More sudden and disastrous is the torrential rain which often accompanies the cyclones prevalent in these seas. The late Charles Hedley has recorded that during the cyclone of January 1918, 55 inches of rain fell at Bowen in three days. The Burdekin and other rivers poured so much fresh water into the sea that a passing vessel drew up buckets of fresh water eight miles from the shore. Few marine organisms and certainly no corals can live in fresh water, and on Stone Island and the adjacent reefs every living thing was killed down to a depth of 10 feet. When Hedley visited this reef six years after, there was still an almost entire absence of animal life; where coral had been there was a thick growth of kelp and other seaweeds. Such seaweeds can normally find little footing on a coral reef, but once established they make it difficult for the coral to return. These are only some of the factors whereby a reef may not only be retarded in its growth, but actually destroyed.

On the credit side are many organisms besides coral, which not only add their own remains to the reef, but help also to bind it into solid rock. Most dead coral is porous and liable to rapid disintegration unless there is something to fill the pores and cement it together. Such an agent is found in a small seaweed called *Lithothamnion* whose tissue is strengthened by a secretion of carbonate of lime. This seaweed grows as an incrustation on dead coral and fills the cracks between different colonies. When it dies its fragments work into the innermost cavities, and, aided by the chemical action of the sea water, bind the whole into solid mass.

This is no new role for calcareous seaweeds. Their remains are found in the coral reefs of the early geological periods, and even before coral reefs existed they seem to have formed whole reefs by themselves. From Pre-Cambrian times beds of limestone show structures which are apparently calcareous seaweeds, and this makes them very early fossils indeed.

Amongst other important organisms are the foraminifera. These are minute unicellular animals which secrete beautiful little shells of carbonate of lime. They live in vast numbers in the sand about the reefs, and their shells are washed into the cracks and hollows. Like *Lithothamnion,* they act not only as a cement, but add by themselves considerably to the mass of the reef. There are many other organisms including shells, the tubes of the sea worms, fragments of the tests of sea urchins, the spicules of sponges, to name but a few, all adding a quota to the growth of the reef.

Thus the struggle goes on day by day, year by year, century by century. The reef as a whole, like its living components, must strive for survival. Under the most favourable conditions it may grow and expand to the limit of its surroundings; it may just hold its own and remain unchanged; or it may lose ground and in the end be completely destroyed. The danger to the Barrier Reef most in mind at the moment is the Crown of Thorns Starfish which consumes living coral polyps. This is a great menace to the reefs, putting an end to the continual growth which maintains the reef from wave attack. A prospective novel danger is the leak of oil from offshore oil wells amongst the reefs if this mineral exploitation were to be permitted! Of course such dangers would destroy reefs as things of beauty for the visitor but it would be a long time in human terms before forces of erosion would destroy a reef even after corals cease to build it.

From what can be seen of existing processes in any one part of the Barrier Reef, it is comparatively easy to reconstruct how they have worked in the immediate past, but the problems of the age and origin of the reef as a whole are more complex. In a length of over 1,200 miles it would be extraordinary if there had never been any variation in conditions, and indeed there are considerable differences, particularly between the extreme northern and southern parts.

The first part of the problem is the nature of the continental shelf, the platform on which the reef is built. Continental shelves are puzzling things in themselves and their origin is obscure. In eastern Australia the shelf extends from low tide to depths of about 300 fathoms, and it varies in width from less than 20 to over 100 miles. Off southern Queensland and New South Wales the edge of the shelf is sharp and defined, and beyond it the sea bottom slopes down at about 5°. This seems little but in fact it is very steep for a submarine slope apart from the practically vertical margins of the coral reefs themselves. This continental slope reaches to the great depth of 14,000 feet. From Rockhampton to New Caledonia there is a submarine ridge, most of it at a considerable depth, but with odd reefs breaking the surface, possibly on the summits of extinct submarine volcanoes. The Coral Sea is not very deep in its western portion. East of Cairns beyond the barrier it drops to 780 fathoms as far as Holmes Reef, which breaks the surface; thence to Divine

Reef it is only from 400 to 500 fathoms deep. Beyond here it slopes down gradually until at about 250 miles from land it is 2,100 fathoms deep.

It is known that in some geological periods land extended beyond the present edge of the shelf; the most logical conclusion is that this edge marks the position of a great fault, and that the ocean bed has sunk at various times and even risen along this line. Why there should be a shelf is most uncertain. Parts of it may have been caused by sea erosion, and a fairly wide belt would be so affected during the fluctuations of sea level in the Great Ice Age; but this belt would be within what is now the 45-fathom level, so that erosion cannot account for the part of the shelf beyond this depth. The best known part of the continental shelf is that off the eastern coast of the United States. The new exploration techniques have shown that this shelf has been built upwards by the accumulation of great thicknesses of sediment from Cretaceous times onwards.

In Queensland, as elsewhere, there is evidence of considerable subsidence in late geological times, the whole of the coastal margin having sunk to a depth of more than 1,000 feet below sea level. Just when the subsidence took place is not known. The general uplift of the highlands is considered to have taken place in late Tertiary times, and it may be that the sinking of the land along the eastern side was a compensatory movement which took place at the same time, or possibly later.

The subsidence complies with the conditions outlined in Darwin's theory of coral reefs, but there is evidence that it was at times too rapid to allow of the orderly evolution from a fringing to a barrier reef. Coral no doubt began to grow from time to time, but where subsidence was more rapid than its growth upwards it was submerged to depths at which it could not live.

It is doubtful if there was a continuous fringing reef along the original Queensland coast, even if the coast formerly extended to where the barrier now is. There is very little coral on the existing coast, since conditions are generally unfavourable. The coastal rivers are numerous, the rainfall is exceptionally high, and it has already been shown how flood water can freshen the surface of the sea miles from land. Rainfall was probably as high in the past as now, in fact in certain phases of the Pleistocene Epoch it may well have been higher. Another factor unfavourable to corals is the mud brought down by rivers. This is carried by currents along the coast, and as it settles on the bottom it smothers any corals which may have begun to grow. Much of the coast is occupied by extensive mangrove swamps where corals can have no place. In fact, it is necessary to go some distance from the shore to reach the zone of coral, and the growth becomes more and more prolific as the shore recedes.

Of particular significance to the history of the reef are three borings, one at Michaelmas Cay near Cairns, the others at Heron Island and Wreck Island in the southern part, and all well inside the outer margin of the reefs. The first bore was carried down to 600 feet, the second to 732 feet, and neither reached bedrock. The Wreck Island bore passed into Tertiary limestones at 530 feet, which continued to the bottom of the bore at 1,898 feet. All passed through coral regarded as Recent in age to 378, 506 and 398 feet respectively and then through sand containing fragments of shell and foraminifera, also at least partly Holocene.

This proves that the floor of the continental shelf is here at various depths greater than 500 feet, and it is probable that at the outer barrier it is much deeper. It also shows that after subsidence the water was at first too deep for coral growth, and that it was not until it had been shallowed by the deposition of sand that the present reefs began to grow.

There is confirmation here that Sir John Murray's theory, that coral reefs commence in waters shallowed by sedimentation, is at least in part true, while at the same time Darwin's theory of subsidence is by no means contradicted. It would seem that all theories, including that of the rising of sea level, have contributed towards the solution of the general problem.

The problem of the origin of the outer reef or true barrier is more difficult of solution. If a bore could be sunk here right to bedrock the information thus gained might well give a clue to the truth. There are several possibilities. The most striking fact is the great resemblance of the outer barrier to the barrier reefs and atolls of the oceanic islands. This would bring it well within the original Darwinian conception, and unless new discoveries contradict this, there is perhaps no need to go further for the solution. Given a slower subsidence of the continental shell in the north than in the south, corals could have found and maintained a footing at an earlier stage, and their upward growth could have kept pace and even exceeded the rate of further subsidence. Corals grow faster on the seaward side of reefs than they do in inshore waters, and this alone tends to raise the outer edge of a reef as a barrier. Not only do the corals grow faster, but they are less liable to sudden destruction from fresh water and mud than those nearer the land, and this again tends to keep the inshore channel open while the barrier becomes more and more consolidated.

The reef is still growing and it still struggles for existence. Much of it has probably grown in the Recent Epoch, that is in the last 10,000 years, though it had its origin long before that. It has also probably not reached its maximum, for while there are open waters at depths of less than 30 fathoms, and while conditions still remain favourable, so long will the coral continue to grow. But the balance between survival and destruction is very small, and the future will no doubt see considerable change.

Australian Caves

Since man began to think, his imagination has been fired by caves. Dark holes leading into the earth were early imbued with mystery. Whither they led no one knew, nor what was hidden in their inner recesses. In days before the scientific study of rock structures it was natural to consider caves as the portals of an unknown subterranean world, with seas and mountains and even cities peopled by strange races or by the gods themselves. Most primitive peoples have viewed caves with dread and avoided approaching their entrances, though a few of the less imaginative have actually made them their homes. A remnant of primitive superstition has survived to the present day, and there are so-called enlightened people who still think of caves as something beyond the border of natural events.

To the writer of fiction caves have ever been a source of inspiration. What would the smuggler or the bushranger be without caves? Jules Verne went further than this. He had his hero go down through a volcanic crater in Iceland to an underground sea full of prehistoric monsters, eventually to reach the surface again in Italy, borne upwards on a raft of petrified wood floating on a stream of molten lava. I doubt if any other writer has quite attained this height of imagination, but poets, who are perhaps entitled to flights of fancy, have also dealt in underground seas. Coleridge wrote of Alph, "the sacred river," that it "ran

Through caverns measureless to man
Down to a sunless sea."

Caves undoubtedly have particular interest and beauty. The deepest caves known at present reach down more than 4,000 feet and the longest have more than 40 miles of passages. The history of many caves goes back well into the Pleistocene and some have their origins in the Tertiary Period.

To speak of caves is to think of limestone, for it is within limestone that most of them are found. There are, nevertheless, other than limestone caves. The commonest of these are hardly worthy of being called caves, and rock shelters is perhaps a better term. Some rock shelters are of moderate depth, but few penetrate far into the rock. They occur most frequently in cliff faces where a soft layer of rock is overlain by a harder one. Seepage and surface weathering cause the softer rock to disintegrate, so that the cliff above is undercut, leaving wide hollows sometimes deep enough to be called caves. The sandstone cliffs about Sydney and the Blue Mountains are full of rock shelters, some of which were used by the aborigines for camps and tribal feasts.

In the far north of Australia the quartzite and conglomerate hills of the Spencer Range about 140 miles east of Darwin are noted for their rock shelters. Inyerluk Hill near Oenpelli is the most famous locality; here the rock shelters are not only deep and spectacular in themselves, but they contain very fine examples of Stone Age painting. Many are very old, and in the folklore of the local aborigines they are the work of mythical creatures still living in the hills. Many of the drawings differ from the usual type at the present day, showing not only the general outlines of fishes, turtles, kangaroos and other animals, but thousands of fine lines depicting the internal anatomy, bones, heart, liver and other organs.

In arid regions wind carries particles of sand and dust and acts as a sand blast, and the softer portions of rocky outcrops may be eaten into. However the hollows so formed are so shallow and gently curved as scarcely to be called caves or even rock shelters. Nevertheless in these dry lands wind does help to form rock shelters by removing the products of weathering. This is true of shallow caves in the granite Musgrave Ranges and in the sandstone Ayers Rock. In unstratified rocks vertical joint cracks have often been enlarged to moderate-sized caves by the action of seepage and weathering. These are generally deeper, higher and narrower than rock shelters in horizontally bedded strata.

More picturesque are caves produced by wave erosion, though again these never penetrate very deeply. Sea caves may be between tide limits or even below them, and as the breakers compress the air at the extreme end of the tunnel they can develop tremendous explosive force. As the sea cuts inward along some line of weakness it may find an outlet on the landward side. It then produces a blowhole similar to that at Kiama on the New South Wales coast, or to others on the rugged Tasman Peninsula in Tasmania. Variations of the same force produce such scenic attractions as Tasman's Arch.

Quite another type of cave is very rarely seen, and is revealed only by chance in mines or deep quarries. These are sealed cavities in the rocks, the open portion of the great faults which once shattered and displaced the strata. One such cavity was brought to light about 60 miles west of Grafton in New South Wales, near the Upper Clarence River. Here lies the ghost town of Lionsville, once a populous mining community, but now lonely and nearly forgotten in a secluded valley on the eastern edge of the tableland.

I visited this cave many years ago. The old hotel was then still in existence, built of solid rough-hewn cedar slabs, and it was possible while lying in bed to read the papers of 30 years before with which the walls had been papered. I believe this hotel has since been burned down, and the mining tunnel leading to the cave has fallen in.

In the days when the Garibaldi mine was yielding good gold, the main drive following the reef had been driven about 400 yards into the hillside. Here the reef was lost, cut off by a great cross fault. The fissure produced by the fault had remained open, forming a cave about 90 feet long and 20 feet wide with a roof about 20 feet high. Deep cracks below filled with water extended to an unknown depth. What made the cave or vugh (to use a mining term) most impressive were the gigantic crystals of calcite lining the sides and roof. These were probably the largest crystals of their kind ever found in the world. Many were six or seven feet across, great obtuse pyramids of shining white. Growing on the larger crystals were numerous small, secondary crystals, exquisite and perfect flat-topped hexagonal prisms, quite transparent, and even in the flickering light of a candle glistening like so many gems.

The largest crystal my companion and I could manage through the narrow drive was about 18 inches across; it now reposes in the Technological Museum, Sydney. The others, I have been informed, were broken up by a German prospector and sent to Germany for making prisms for petrological microscopes and other optical instruments. This variety of calcite, noted for its transparency and strong double refraction, is known as Iceland spar and is quite valuable. It is sad to think that this wonder cave has now disappeared for ever. Should such another ever be found it is hoped that some action will be taken to preserve it for posterity.

There are other types of caves, some of which have been found in Australia. In regions of recent volcanic activity lava-flows have cooled and hardened on the surface while the still molten part has flowed away from beneath, leaving extensive caverns and passages. There are remnants of some of these caves in parts of Queensland. West of the Burdekin River in the rough basaltic country north of Charters Towers deep pits are often met with, sometimes filled with water, but sometimes with openings below leading to subterranean passages. These were occasionally used by the blacks as hiding-places. The caves all occur within old lava-flows, and the pits are where the original roof of lava has caved in. It is probable that there are many more such caves hidden beneath the surface in the district.

The foregoing are minor types of caves. It is limestone caves that are not only the commonest and largest, but from the scenic point of view the most attractive and popular. Wherever there is limestone, caves are likely to be found, small or large.

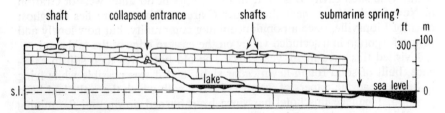

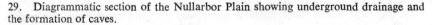

29. Diagrammatic section of the Nullarbor Plain showing underground drainage and the formation of caves.

All limestone caves owe their origin to the same causes, though there is considerable difference in type. Limestone is composed of calcium carbonate, which is slowly soluble in water charged with carbon dioxide. Rainwater absorbs this gas from the atmosphere, dissolves the limestone, and carries it away in solution. The rain also finds its way into cracks and crevices within the limestone and dissolves out large cavities and caves. Even more important than the carbon dioxide of the atmosphere is that provided in soil and plant litter on the ground. Root respiration and microbiological activity, which rots soil organic matter, together enrich soil air with carbon dioxide to levels much higher than are found in the atmosphere we breathe. So when rain water percolates through the soil, it also acquires much more carbon dioxide in solution. As a result we find much greater contents of calcium carbonate in waters coming from limestone than rainwater could dissolve. The availability of this biological carbon dioxide to make natural waters acid is one of the factors governing the development of caves. Locally other acids can be important—sulphuric acid from the weathering of certain minerals and organic acids from the rotting of vegetable matter. The size and form of the caves also depend upon the structure of the limestone formation and its relation to the surrounding rocks.

Two distinct types of limestone country are to be found, in each of which there may be caves. One is exemplified in the caves of the Nullarbor Plain, the other in those of eastern Australia. Limestones of the later geological formations are generally not distorted by folding, and often lie horizontally bedded over considerable areas of country, forming flat tablelands at a varying height above sea level. Drainage from such areas tends gradually to find its way underground, dissolving channels along minor cracks and fissures until even the main streams disappear from the surface. The water eventually emerges through springs or from the mouths of caverns at the base of the tableland. Such a bed of limestone may be literally honeycombed by caves, and though apparently solid upon the surface, at this stage it is well upon the way to its final dissolution.

Such formations are not common in Australia, where most of the limestones belong to very early geological periods and have been folded and contorted in conjunction with the underlying and overlying strata. Nevertheless, some do exist, and striking examples occur where the Nullarbor Plain abuts on the Great Australian Bight in both South and Western Australia. The limestone forming the Nullarbor Plain was laid down under the sea in the Tertiary Period as the land sank below sea-level over large sections of southern Australia. The sea penetrated over 150 miles beyond the present coast, and when the sea bed rose again it carried with it a great thickness of limestone. This stands now as a plain 150 to 500 feet high and many hundreds of miles wide from east to west.

Throughout the whole of this great area the limestone is now honeycombed with caves. It is probable that only a few of them are known, and fewer still have been explored. How many more are hidden beneath the surface it is impossible to surmise. That there are many is obvious, for over the whole of the Nullarbor Plain the drainage is underground. There is on the surface neither river nor creek. From the surrounding country a few former river courses, never followed today by water, can be traced from the north into the plain. They did not reach across the whole of it, ending north of the railway or about its latitude. The present climate is dry, the annual rainfall being from 15 to under 10 inches; but this was land through the whole of the Pleistocene Epoch, when there were phases of more effective precipitation and runoff. Even now, when the rain does fall it may be quite heavy for a brief period; yet there is no run-off, and every drop that is not quickly evaporated sinks into the ground.

Below the surface the water has gradually dissolved channels and galleries, some of great extent. Most of the caves have very small entrances, and unless special search is made they can be easily overlooked. Some larger entrances do exist, but only where large sections of the roofs have fallen in.

The presence of caves below the Nullarbor Plain has been known for a long time to both aborigines and white men, but much of the credit for recent discovery and exploration must go to Captain John Thomson of the Adelaide Pilot Service, who has made this his hobby. Captain Thomson in one brief flight discovered the entrances to no less than 42 hitherto unknown caves. Access to many of those he explored was by rope or ladder down narrow vertical shafts, and galleries extended from the bottoms of these. Just across the Western Australian border is Weebubbie Cave, where is virtually an underground lake. This was explored in a canoe to the end of a passage with a flat roof 40 feet above the water. Koonalda Cave in the South Australian part has a great dome-shaped chamber between two lakes, the roof of which is 220 feet above the water, though there is a pile of rock 115 feet high immediately below the dome.

Just where the water from the caves finds its outlet to the sea is not certain. The plain continues to the coast, where it terminates in the high escarpment of the Hampton Range. It is probable that most of the springs which must somewhere exist emerge below the level of the sea itself.

The Nullarbor Caves, unlike most limestone caves, are rarely embellished with formations of stalactites and stalagmites. The limestone is quite pure and there are joints in the roofs of many caves which yet lack stalactites. The explanation rests in poverty in water; there is little water dripping today nor can there have been in the past. Nevertheless, the changing colour of the different layers of limestone, the impressive spaciousness of the lofty halls and the sombre mystery of underground waters give the Nullarbor Caves a beauty peculiarly their own.

Of similar origin are the Naracoorte Caves in South Australia. These are just across the Victorian border, about seven miles south of the township of Naracoorte. There is little in the surface scenery to suggest spacious caves underground. The surrounding country is swampy and sandy and is covered with stringybark. It is a low plateau of Tertiary marine rocks, part of the great formation extending northward into New South Wales, and disappearing south and east under the newer volcanic rocks of Mt Gambier and the western plains of Victoria.

Ten caves have been opened to the public. Most have been found by accident, for the entrances are small surface holes, which lead, nevertheless, to extensive galleries and chambers. Many undiscovered caves probably exist beneath the country both to the south and east. Some of the known caves are very extensive. The Big Cave, quite close to the entrance, is 200 feet long, 70 feet wide and 20 feet high. One feature distinguishing them from the Nullarbor Caves is the abundance of stalactites and stalagmites decorating the roofs and floor. These are very beautiful and comparable with those of the older limestone formations in the eastern highlands.

Along the whole west coast of Western Australia the formation of the coastal limestone contains numerous caves. Most are small, but there are some known caves of considerable extent. Some conform to the type of the Nullarbor Caves and have been formed by surface water percolating through the porous limestones. Others differ in that streams from impervious rocks to the east flow into the limestone to play a prime role in forming caves such as the Mammoth Cave near Margaret River. Apart from its beauty, and there are many fine stalactite formations, a noteworthy feature of Mammoth Cave is the great number of marsupial bones accumulated beneath its floor. Between 1905 and 1910 systematic collections were made, and there were discovered the remains of many extinct species, including the giant diprotodon, the nototherium, kangaroos, wallabies and a skull of the Tasmanian devil, extinct upon the mainland but still living in Tasmania. Some of the fossils are known to be more than 31,000 years old. Many new caves have been found in recent years, particularly between Cape Naturaliste and Cape Leeuwin, several of which are as impressive as Mammoth Cave. One of these with many remarkable features is Jewel Cave near Augusta, which had been opened for tourists.

The limestone caves familiar to the inhabitants of the eastern States are of a different type. They are found where rocks of the older geological formations have been so folded that they stand on end or dip at very high angles. Where the land has been elevated in later geological times, and where rivers still excavate deep valleys and gorges, harder strata of rock lying across the valleys form barriers and slow down erosion higher up the streams. Waterfalls, cascades and rapids develop, but where the barrier is limestone, the river may find its way underground and form caves. The older limestones of eastern Australia are generally hard, compact and resistant to erosion, though like all limestones they may be dissolved by water charged with carbon dioxide.

When the eastern highlands of Australia were elevated in the Tertiary Period conditions for the formation of caves became favourable in many localities where there were beds of Silurian and Devonian limestone. The diagrammatic section in Figure 30 shows just how such caves are produced.

In Figure 30, a stream has excavated a valley, and crossing its course is a hard belt of limestone tilted almost vertically and interbedded with soft shale and other rocks. The bed of the stream was lowered while the harder limestone remained as a barrier across its course. This barrier was only temporary, for the water eventually found its way through cracks and crevices and dissolved channels for itself. As the bed of the valley became lower, the underground river itself sought lower levels, leaving above it a series of passages through which it formerly flowed.

Many the eastern caves conform to this pattern, though there is variation in detail in different localities. The extent of the caves depends on the thickness of the limestone and the angle it makes with the course of the stream, also the volume of water and its corrosive power. Extensive beds of limestone may exist near the confluence of two or three streams, as at the Wombeyan Caves in New South Wales. Here is more than one underground river, and the whole system of caves is very complicated. At Borenore the limestone barrier is so corroded that it forms a single arch high above the stream.

The last stage is reached when the river has cut its bed so low that it can excavate no deeper. The valley widens, the hills on either side become lower and rounded, and alluvium forms flats upon the valley floor. The limestone barrier itself is slowly eaten away, and only its remnants are left at the sides of the valley. These may contain a few small passages, all that is left of a large series of caves. At Moor Creek near Tamworth in New South Wales there is an excellent example of such a series of caves in the last stage of their existence. Moor Creek now meanders through a broad valley towards the Namoi River, where once it descended in a narrow gorge from the Moonbi Ranges above. A large bed of limestone once lay right across its course, but all that now remains is a limestone bluff on the northern side, in which there are a few minor caves.

The decoration of caves with stalactites and stalagmites is another phase of the solution of carbonate of lime in water and its subsequent deposition. In nearly all caves, in addition to the main channels, there are cracks in the roof through which water seeps. When a drop of water remains suspended for a while before it falls to the floor, it may diffuse some carbon dioxide to the cave air if it has a greater concentration, which is often the case; as a result, it parts with some or all of the dissolved carbonate of lime. This remains as a thin ring of solid matter. In the course of time, as ring is deposited on ring, the long pendant stalactites grow downwards. Where the seepage is continuous along a crack the stalactites fuse together into translucent shawls, one of the most beautiful of formations.

Stalagmites grow upwards from the floor, often meeting the stalactites to form complete pillars. They are produced similarly, but from water dripping from above, which diffuses more carbon dioxide to the cave air and parts with some of its solid matter before it trickles away. Helictites or "mysteries" are peculiar formations, and as yet there is no complete explanation of their origin. They are usually long and slender, and grow outwards and even upwards from the sides of stalactites, often twisting into a great variety of grotesque shapes. When broken they are seen to be composed of crystalline calcite, but there is no central hollow as in the stalactites. They too have obviously been deposited from solution, but the water is forced through hairline cracks or microscopic pores in the calcite in very small amounts. There is never enough water emerging for gravity to overcome surface

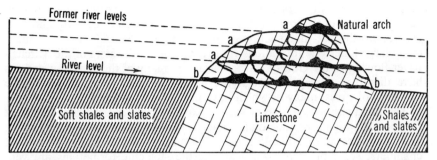

30. Diagrammatic section showing how caves are produced when a steeply dipping limestone bed forms a barrier across a river valley. (a) Entrances to caves on former river levels; (b) entrance to present underground river.

tension and to get control of water movement, and so of the form of the crystal being precipitated.

In small hollows, where water accumulates and loses carbon dioxide to the air, the carbonate of lime may be deposited as single small crystals or groups of crystals of the mineral calcite, delicate prisms or faceted pyramids which sparkle like gems in the light of a candle or torch. Though the crystals of one mineral may assume a great variety of shapes, their molecular structure is so constant that all the shapes conform to a definite mathematical pattern or system. In calcite the system is hexagonal, and in all the complications the hexagonal symmetry is always present. Basins of crystals like baskets of gems are probably the most beautiful of cave formations, but are less common than stalactites and stalagmites.

There is much misconception about the rate of growth of stalactites. The rate naturally varies according to many factors, the rate at which the water drips, the amount of solid it contains in solution and the rate of carbon dioxide diffusion to the air. The estimates of millions of years often quoted by guides can be very misleading. Under favourable conditions growth is comparatively rapid. In the Moor Creek Caves some stalactites had been broken off by a team of visiting footballers some 16 years before my visit, and small new stalactites up to two inches in length had already formed upon the broken stumps. On a large stalactite growth might be continuous over the whole surface if it is constantly wet by seepage. Where seepage is small and intermittent, growth, on the other hand, might be very slow. On this basis the age of large stalactites might well run into thousands of years, but hardly the millions so often glibly averred. Radiocarbon dating of stalagmites has so far shown that even large decorations of this type are to be thought of as tens of thousands of years old only.

In Queensland, amongst the many limestone caves, the Chillagoe Caves have long been renowned for their extent and beauty. Across the main divide in the far north and to the west of the basalt-capped Atherton Tableland lies a complex geological tangle of formations. Here great intrusions of granite have penetrated into the twisted and folded strata of the older periods. In the heart of this country lies the formerly rich mining township of Chillagoe. Just to the west of the town is an extensive belt of Silurian rocks running in a general north-west and south-east direction. The Silurian rocks contain many large limestone formations. This forms the headwaters of the Mitchell River flowing to the Gulf of Carpentaria, and many of the tributary streams have dissolved passages through the limestone and produced a veritable labyrinth of caves, many of which have never been explored.

The main entrance to one of the best known caves is through a fissure at the foot of a high limestone bluff about three miles to the west of the town. This gives access to a whole series of winding galleries and high vaulted chambers, some of great size. Many of the stalactite formations are very beautiful, but unfortunately there has never been proper control, and those in the more accessible caves have been badly damaged by vandals. Even more extensive are the caves at Mungana, now a deserted town about 10 miles north-west of Chillagoe. The full extent of the Mungana Caves is unknown. Recent careful explorations of many caves have failed to reveal any running streams, though their lower parts must develop pools in the wet season. Nor have any upper galleries representing former stream passages been mapped as yet.

Apart from the caves the external weathering of the limestone has produced much rugged and peculiar scenery in the district. The bulk of the rock is compact and splintery, and the tropical downpours have corroded the surface of individual rock masses into series of deep vertical furrows. Many of the outcrops have been corroded into grotesque forms. Local names given to rock formations are often self-explanatory. There is a huge pile of limestone called "The Tower of London", another is "The Dome", or "Elephant Rock"; then there are the "Polar Bear Rock, "Tank Rock", "Balancing Rock", and many others.

Though the Jenolan Caves are the most famous in New South Wales, they are not the earliest known, for the Yarrangobilly Caves were known to the early settlers at least ten years before. The Jenolan Caves were actually discovered by the bushranger McKeown, who found a refuge in one and made it his headquarters. It was in 1841 that a posse searching for McKeown discovered his hideout and revealed the wonders of the caves to the outside world.

So much has been written about their stalactites and stalagmites, their grottoes and chambers; there have been so many comparisons to cathedrals, altars and tapestries; there have been so many fancied resemblances to birds, animals, statues and other objects, that it is not necessary here to describe individual caves in detail. Guide-books compete in adjectives, but there is a marked similarity in the interor decorations of caves, whether they are at Jenolan, Buchan or Naracoorte. We are here primarily concerned with causes, and the Jenolan Caves are typical of most of the limestone caves in eastern Australia.

They lie in a deep gorge in the rough country beyond the Blue Mountains, and just below the eastern slope of the main divide. Here a thick belt of

Silurian limestone is interbedded with softer slates, and in the period of erosion following the last elevation the limestone has remained above general water level while the slates have been cut away.

Features of the Jenolan Caves are the great natural arches at former river levels, where the streams once flowed through the limestone barrier. Of these Carlotta Arch is now high above the existing river level, but lower down, the road to Cave House actually passes through the Grand Arch, a colossal chamber 40 to 70 feet high, 450 feet long and from 35 to 80 feet wide. The Devil's Coachhouse is even larger, its dome towering to a maximum of 160 feet above the floor, while its length is over 400 feet and its width 120 feet.

The Wombeyan Caves, 47 miles north of Goulburn, are similar in structure to the Jenolan Caves, though their immediate surroundings are neither so steep nor so rugged, and the surface features of the caves are not so prominent. The belt of limestone is about one mile square, and there is quite a complicated pattern of chambers and passages on different levels, with several underground streams all playing their part in cave production.

In the southern part of New South Wales there is a great development of Silurian and Middle Devonian marine rocks. They form much of the tableland and all are greatly folded, with a number of small belts of limestone. Caves are therefore found in many localities, and some are as yet incompletely explored. The Bungonia Caves east of Goulburn, unlike those at Jenolan, do not lie in a gorge, but are right on the tableland, on the edge of the "Look Down Precipice", 1,000 feet above Bungonia Creek. As might be expected, the caves here are very deep, forming shafts rather than corridors. The deepest surveyed cave on the mainland of Australia, Odyssey Cave, with a depth of 485 feet, is found here. Another cave called the Fossil Cave-Hogans Hole Extension has a long nearly horizontal passage over 3,000 feet long but it is exceptional for the area.

The famous Yarrangobilly Caves illustrate another variation in the relation of a limestone belt to the surrounding country. They lie at an elevation of 3,000 feet on the southern tableland between Tumut and Kiandra, just where it begins to rise towards its greatest height on the Kosciusko Tableland. The Yarrangobilly River here flows south through the little village of Yarrangobilly, and is flanked below the town for some six miles by high cliffs of limestone. The limestone belt is less than a mile wide and runs parallel to the river, forming a plateau above the cliffs, where it is capped in places by later flows of basalt. Along the river itself the limestone has been mostly removed by erosion, except in one place where it forms a ridge across the valley. Beneath this ridge the river flows under a low arch. Behind the limestone plateau numerous small tributary watercourses flow west towards the main river, but all disappear at the edge of the limestone belt, flowing underground through a network of channels before they emerge through the face of the flanking cliffs (see Figure 31). The small tableland is literally honeycombed with caves, and it is probable that before the river removed the main barrier of limestone the cave system was even more extensive than it is now.

The most famous caves in Victoria are the Buchan Caves, situated in the

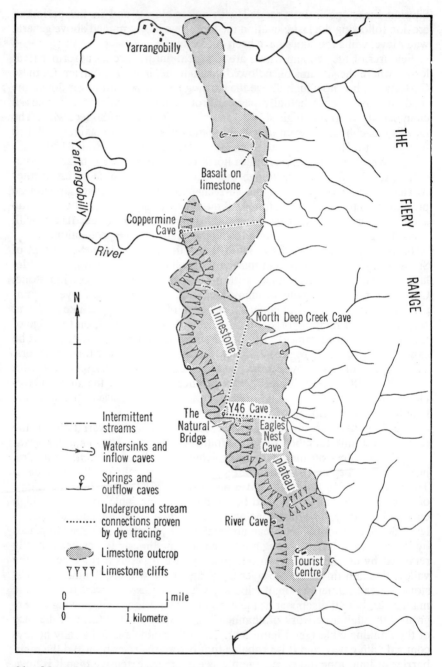

31. Map of the Yarrangobilly limestone belt showing some caves and underground river connexions.

mountainous country north of the Gippsland Lakes. The word Buchan, though suggestive of Scotland, is really an aboriginal term meaning dilly-bag, and is more correctly spelt "bukken". The caves are of considerable extent, and many are of great beauty. Many are only partly explored and are not yet open to the public. The story of their origin is very similar to that of other caves in the eastern highlands. The Buchan limestones are of Middle Devonian age, and are rich in marine fossils of that period. They lie in a broad fertile valley above a series of hard porphyries into which they have been down-faulted, so that surrounding them is a ring of forest-clad porphyry ranges.

Tasmanian limestone caves have been mostly found in the foothills of the main tableland both in the north and the south. This tableland, it may be remembered, is composed mainly of great sheets of dolerite, but outcropping on the margin are rocks of the Ordovician and Silurian Periods, in which are limestones of considerable extent. The high rainfall of south-western Tasmania is favourable to cave formation but many outcrops are in the flat bottoms of valleys, and caves there are liable to be full of water and so difficult to explore. However on Mt Anne, dolomite, another soluble carbonate rock, reaches to high levels and deep caves have already been discovered there after little exploration.

The best-known Tasmanian caves are near Mole Creek behind the north coast. Here King Solomon's Cave has many beautiful stalactite formations and some wonderful examples of the mysterious helictites. Near by one cave has an underground stream more than two miles long. The Marakoopa Cave is in the same district. In the south the Newdegate Cave is the best known. It is near Hastings, 70 miles from Hobart, in the foothills of Adamson's Peak. Further south is Australia's longest cave. This is Exit Cave in thick bush country of high rainfall near Ida Bay. There are about ten miles of passage known of which seven have been surveyed and there is a shaft 720 feet deep leading down to the cave from the ridge above. However a deeper cave over 900 feet deep has recently been descended on the southern side of Mt Field National Park near Maydena.

Scenery—A National Heritage

In the foregoing chapters a number of the scenic phenomena of Australia have been dealt with, not only because of their aesthetic appeal, but also because of the age-old story behind them. The beholder may be awe-struck and exhilarated at mighty mountain ranges and gorges, but the quiet rustic scene can convey no less a wonder at the slow unceasing forces for ever reshaping the face of the earth. The hills, which would seem to be eternal, have been shown to be the most ephemeral of things and mere episodes in geological history. It is often the humbler features of landscape, the plain that is nearly at sea level or the low plateau, that may survive from age to age, while elsewhere mountains rise and disappear. The Barkly Tableland and the hinterland of Western Australia have lain much as they are since the dawn of geological time, while on the eastern side of Australia lofty mountains have reached up into the clouds not once but many times, only to be worn slowly to low levels again.

Scenery may thus have a twofold interest. When the eye is satiated with the beauty of a landscape, even when the view is lacking in colour and form, the mind may find as great an appeal by seeking the story that lies beneath. There is ever a fascination in interpreting the present in terms of the past. From evidence gathered in cuttings, quarries and mines the former scene is reconstructed, and the dimensions of space and time dwindle until they come within the bounds of conception. There is no fragments of landscape that has not within it the story of an illimitable past, or forces that have acted age after age, until finally its present shape has been assumed. To those possessed of any knowledge of the causes of scenery there can be no boredom in travel, for everywhere there are evidences to be perceived, facts to be learned and problems to be solved.

When the scenes of a whole continent are discussed separately there must of necessity be some repetition. In each district the same forces have been at work, and many landscapes, however dissimilar, represent different stages in a similar sequence of events. This suggests a broad classification of Australian scenery into three groups, the old, the intermediate and the new.

The old scenery is that which was shaped in the early geological periods, thenceforth to remain virtually unchanged. In the early chapters such ancient lands as Yilgarnia were described as surviving from times as far back as the

Archaeozoic Era 2,500 million years ago. There was Nullagine in Western Australia, perhaps 1,800 million years old, and the Barkly Tableland, part of which has remained land for 500 million years. Other fragments of ancient land are to be found at Broken Hill, and possibly in Tasmania.

Lands of intermediate age are sometimes in the final stages of dissolution. Such are the Macdonnell Ranges in central Australia, which were last under the sea in the Ordovician Period. However there is still some vigorous relief here which will take a long time to be reduced to a plain in the prevailing arid climate. Though of much later age, the plains of north-western New South Wales and central Queensland might also be considered here. This area has remained unaltered since the drying of the great lakes at the end of the Cretaceous Period, and is apparently so stable that it may well remain as it is for long ages to come.

New scenery is that which has had its main outlines determined in comparatively recent times and which is still being shaped and altered by erosion. The highlands of eastern Australia come under this heading, as do certain portions of the coast. The main features were determined in the Tertiary Period, when elevation lifted the tableland to heights from 2,000 to 7,000 feet above the level of the sea; the details are still being modified by the action of frost, rain and wind.

We have seen how different geological formations have influenced the present shape of the mountains, how the horizontally bedded sandstone capping the Blue Mountains has left long flat-topped ridges separated by deep abysses, while in the Victorian Grampians hard sandstone dipping at an angle has produced quite a different contour. We saw in the Queensland corridors how great faults raised large block mountains above the general level and at the same time depressed rift valleys sometimes below the level of the sea. We saw how the falling and rising of the sea in the Great Ice Age modified the shape of the coast, drowning river valleys and separating islands from the shore. In Tasmania alone we saw how the Ice Age had a direct effect upon the land, covering much of the island with a vast sheet of ice that left tangible evidence of its passing in the form of cirque mountains and innumerable lakes. Again on the mainland we saw how volcanoes built up other groups of mountains, the Canobolas, the Warrumbungles, the Nandewars and the Glass House Mountains. And finally we discussed the somewhat special scenic phenomena, the Great Barrier Reef and limestone caves.

Having done these things, there remains still a brief review of the interaction between civilized man and the scenery amidst which he lives. There is growing evidence that Australians are becoming more scenically conscious. Popular magazines and journals are devoting greater space to Australian travel and exploration, and more and more books are being published to throw light, not only on the geography, but also on the geology and natural history of little-known areas. Clubs of bushwalkers are finding new paths in wild country close to the cities, and enthusiasts are organizing serious expeditions into many hitherto inaccessible parts.

Modern facility for travel is perhaps the greatest incentive to this growing curiosity about what lies in the innermost recesses of the continent. Older members of the present generation well remember the days before motor cars were in general use, when roads were fewer and rougher, and certainly before the aeroplane reduced a journey of days or even weeks to a few hours. It was not *then* possible for a party of schoolboys to spend a vacation in exploring Ayers Rock, one of the strange mountains of central Australia. All this is much to be desired, for it means that in the near future the blank spaces on the map will become smaller and smaller. With growing knowledge will come also the realization of what a really wonderful country Australia is.

The aeroplane is not only revolutionizing exploration and surveying, it is giving a new and instantaneous perspective to large scenic features. Such curious formations as Wilpena Pound in South Australia and the Waterhouse Range in central Australia formerly could be examined only piecemeal from the ground, and it took long patient geological survey to reveal their general structure. It is now possible to view them from the air, and, as shown in aerial pictures in this book, such unique shapes and interesting structures can be seen at a glance. In the wild west country of Tasmania and over the waterless deserts of the west the aeroplane is doing in a few hours what previously took months of arduous travel and toil.

As means of communication lessen distance, as outlying parts are drawn closer and closer together, so must one other problem surely arise. That is how to preserve for future generations this heritage of beauty and scientific interest. It might be thought that scenery is on such a scale that it cannot be artificially destroyed. It is true that while man may level a hill when quarrying for minerals, or dam a valley and create a lake, he cannot yet remove whole mountains or build others in their place. He may nevertheless do irreparable damages to the details of the whole picture. He may also, and sometimes quite unwittingly, destroy much that is of scientific value.

Vandalism is due as much to ignorance and thoughtfulness as to a sole love of destruction, and those who despoiled the Chillagoe caves of their stalactites were simply selfish and heedless of the rights of those who came after them. At times no one particularly is to blame. The unique cave of crystals at Lionsville is for ever lost to the public because it was no one's business to see that it was preserved.

In the path of progress some destruction is inevitable and at times necessary for economic reasons. Thus this is the case made for the drowning of the Lake Pedder area in south-west Tasmania. Efforts are being made to preserve the Macquarie marshes in western New South Wales, otherwise they would have dried up on the completion of dams higher up the Macquarie River. As the marshes are one of the few breeding-places of the ibis and other valuable aquatic birds, their disappearance would have been a great disaster. The Lamington National Park in Queensland and the Scenic Reserve in central Tasmania are two of a number of areas officially set aside for preservation in a natural state. It is hoped that in the future similar reserves will be proclaimed in the far outback to keep intact for posterity some of the marvels of the Australian hinterland.

National parks are priceless assets for the future, but even outside these sanctuaries there is great need for conservation. As settlement has spread there have been much needless cutting down of forests, overstocking of pastures and senseless killing of unique birds and marsupials. Bushfires lit by careless hands have wrought untold damage to the native flora and fauna. As plants and animals are an integral part of the scenery a plea for their preservation is not here out of place. The story of the decimation of Australian animal life is a sad one, several species have become extinct and others are in extreme danger. For such unique animals as the koala and the platypus, for such birds as the ibis, the lyre bird and the black swan, the public conscience is beginning to awaken, but there are many rare plants and animals whose destruction is more by accident than design.

There is, for instance, a cypress pine, *Callitris muelleri,* known from only two localities, one on the cliff edge at Wentworth Falls, the other on the hillside just east of The Spit, Middle Harbour, Port Jackson. The small patch at the latter locality was destroyed by the construction of the tramway many years ago, the other at Wentworth Falls may still possibly be in existence. Another rare plant, *Boronia thujona,* has also been recorded from only a few localities, one at Bundanoon on the southern tableland, another near Narrabeen at the head of Middle Creek, with another small patch near by at Warriewood. The patch at Bundanoon was destroyed by a bushfire, and as Warriewood is now closely settled, it is to be feared that this too has gone.

These are but two instances of destruction, not caused by wantonness, but by accident or from lack of knowledge. To prevent such happenings it is not merely sufficient that people have no desire to destroy; there must also be the positive approach, the will to preserve, and this too should be combined with possession of the necessary scientific information.

Even less apparent than the destruction of rare living plants and animals is that of fossils found within the earth. Fossils are the remains of once living things buried in the rocks, and from them is learned, not only the history of life on the earth, but the story of geographical changes leading up to the present scene. Good fossils are rare and their preservation is due to the billion-to-one chance. Having been so preserved for millions of years, it is tragic that they are finally lost through carelessness or lack of recognition. Many are inevitably destroyed by erosion and the natural dissolution of rocks; but others from quarries, cuttings and mines are easily obtained and should be saved. In America certain areas famous for their fossils have been declared reserves, and only accredited scientific workers are granted permits to dig in these places.

In the brickpits at Brookvale near Sydney it was the interest of an intelligent foreman and a quarryman that preserved many unique fossils of fish, insects, plants and curious extinct amphibians that added much to the knowledge of life in the Triassic Period. The number of such valuable specimens that elsewhere have vanished in the making of bricks or have been thrown on mullock heaps of mines will never be known.

It is only true patriotism that can halt destruction, that can preserve intact for posterity the thousand and one things that characterize the face of the country. If patriotism is defined as a love of country it goes further than pride in its human achievements, whether in commerce, manufactures, art, literature or even war. It involves a love of the land itself, for out of the land is evolved much of the national character. Hills, plains and forests all leave their imprint on the human mind, as do the sun, the wind and the rain, the golden sands of the beaches and the solitudes of the far outback. An understanding of nature is a stimulus to true patriotism, for it induces not only love of an environment, but humility when contemplating the stupendous power of the forces which created it. Let us therefore not only love our country, but let us consider this natural wonderland of Australian as a sacred heritage, something to be preserved unsullied and intact for the benefit of generations to come.

Index of Principal Place Names